Keith J. Devlin

Fundamentals of Contemporary Set Theory

Springer-Verlag
New York Heidelberg Berlin

Keith J. Devlin
Reader in Mathematics
University of Lancaster
Lancaster, England

AMS Subject Classification: (1980) 04-01

Library of Congress Cataloging in Publication Data

Devlin, Keith J
Fundamentals of contemporary set theory.

(Universitext)
1. Set theory. I. Title.
QA248.D38 511'.3 79-17759

Printed in the United States of America.

9 8 7 6 5 4 3 2 1

ISBN 0-387-90441-7 Springer-Verlag New York Heidelberg Berlin
ISBN 3-540-90441-7 Springer-Verlag Berlin Heidelberg New York

Preface

This book is intended to provide an account of those parts of contemporary set theory which are of direct relevance to other areas of pure mathematics. The intended reader is either an advanced level undergraduate, or a beginning graduate student in mathematics, or else an accomplished mathematician who desires or needs a familiarity with modern set theory. The book is written in a fairly easy going style, with a minimum of formalism (a format characteristic of contemporary set theory).

In Chapter I the basic principles of set theory are developed in a "naive" manner. Here the notions of "set", "union", "intersection", "power set", "relation", "function" etc. are defined and discussed. One assumption in writing this chapter has been that whereas the reader may have met all of these concepts before, and be familiar with their usage, he may not have considered the various notions as forming part of the continuous development of a pure subject (namely set theory). Consequently, our development is at the same time rigorous and fast.

Chapter II develops the theory of sets proper. Starting with the naive set theory of Chapter I, we begin by asking the question "What is a set?" Attempts to give a rigorous answer lead naturally to the axioms of set theory introduced by Zermelo and Fraenkel, which is the system taken as basic in this book. (Zermelo-Fraenkel set theory is in fact the system now accepted in "contemporary set theory".) Great emphasis is placed on the evolution of the axioms as "inevitable" results of an analysis of a highly intuitive notion. For, although set theory has to be developed as an axiomatic theory, occupying as it does a well established foundational position in mathematics, the axioms themselves must be "natural" : otherwise we would be reducing everything to a meaningless game with prescribed rules. After developing the axioms, we go on to discuss the recursion principle -

which plays a central role in the development of set theory, but is nevertheless still widely misunderstood, and rarely fully appreciated - and the Axiom of Choice, where we prove all of the usual variants such as Zorn's lemma.

Chapter III deals with the two basic number systems, the ordinal numbers and the cardinal numbers. The arithmetics of both systems are developed sufficiently to allow for most applications outside set theory.

In Chapter IV we delve into the subject set theory itself. Since contemporary set theory is a very large subject, this foray is of necessity very restricted. We have two aims in including it. Firstly, it provides good examples of the previous theory. And secondly, it gives the reader some idea of the flavour of at least some parts of pure set theory.

Chapter V presents us with a modification of Zermelo-Fraenkel set theory. The Zermelo-Fraenkel system has a major defect as a foundational subject. Many easily formulated problems cannot be solved in the system. The Axiom of Constructibility is an axiom which, when added to the Zermelo-Fraenkel system, eliminates most, if not all, of these undecidable problems.

Finally, in Chapter VI we give an account of the method by which one can *prove* within the Zermelo-Fraenkel system that various statements are themselves *not provable* in that system.

Both Chapter V and Chapter VI are non rigorous. Our aim is to *explain* rather than *develop*. They are included because of their relevance to other areas of mathematics. A detailed investigation of these topics would double the length of this book at the very least, and as such is the realm of the set theorist. (Though we would, of course, be delighted to think that any of our readers would be encouraged to go further into these matters.)

Chapters I through III contain numerous easy exercises. In chapters I and II they are formally designated as "Exercises", and are intended for solution as the reader proceeds. The aim is to provide enough material to assist the reader in understanding fully the concepts which are introduced. In Chapter III the exercises take the form of simple proofs of basic lemmas, which are left to the reader to provide. Again, the aim is to assist the reader's comprehension.

At the end of each of Chapters I through III, there is also a small selection of problems. These are more challenging than the exercises, and constitute digressions from, or extensions of the main development. In some instances the reader may need to seek assistance in order to do these problems. The general format is as follows. Each problem commences with some theory. Then follow various assertions indexed by capital letters (A), (B), (C), etc. It is these assertions which the reader should attempt to prove.

Chapters IV through VI provide little more than a brief glimpse into the fascinating set theoretical world beyond the boundaries set in the earlier chapters, and are not supplied with exercises. Only the set theorist needs to achieve any significant degree of competence in these areas, and for him there are other books available.

The basis for this book is a series of lectures I gave at the University of Bonn, West Germany, in the years 1975 and 1976. My stay in Bonn was supported by the Deutscheforschungsgemeinschaft. In addition to acknowledging their generous support, I should also thank the students in Bonn, who so courteously listened to my appalling German for weeks on end. The first book I ever read on set theory was Rotman and Kneebone [6], and no doubt the influence of their book is apparent in mine.

Chapters are numbered by Roman numerals. Each chapter is divided into sections numbered by Arabic numerals. The n'th lemma or theorem in section m is denoted as m.n within the same chapter, and prefixed by the Chapter number otherwise. The symbol $\square$ is used to indicate the end of a proof.

Finally, I wish to thank Paddy Farrington for reading the final manuscript and pointing out various mistakes, and Sylvia Brennan for the excellent typing.

Contents

Chapter I. Naive Set Theory

Before we commence, let us make one thing clear: in this book we shall only study one *set theory*[†] ; and that theory is a rigorous theory, based upon a precise set of axioms. However, it is possible to develop the theory of sets a considerable way without any knowledge of those axioms. And, indeed, the axioms can only be fully understood after the theory has been investigated sufficiently. This state of affairs is to be expected. The concept of a "set of objects" is a very intuitive one, and, with care, considerable, sound progress may be made on the basis of this intuition alone. Then, by analysing the nature of the "set" concept on the basis of that progress, the axioms may be "discovered" in a perfectly natural manner. Following Halmos [3] we refer to the initial, intuitive development as "naive set theory". A more descriptive, though less concise title would be "set theory from the naive viewpoint", with perhaps a parenthesised definite article preceding "set theory". Once the axioms have been introduced, this "naive set theory" can be re-read, *without any changes being necessary*, as the elementary development of *axiomatic set theory*. In other words, there is just *set theory*. But now, down to business.

[†]This is not strictly accurate, since our discussion of the axiom of constructibility in Chapter 5 really amounts to a consideration of a set theory other than the Zermelo-Fraenkel set theory, which forms the main topic of the book. But for the present discussion, this point is more misleading than helpful, which is why it has been relegated to a footnote.

1. What is a Set?

In naive set theory we assume the existence of some given domain of "objects" out of which we may build sets. Just what these objects are is of no interest to us. Our only concern is the behaviour of the "set" concept. (This is, of course, a very common situation in mathematics. In algebra, say, when we discuss a group, we are (usually) not interested in what the elements of the group are, but rather in the way the group operation acts upon those elements.) When we come to develop our set theory axiomatically we shall, in fact, remove the necessity of an initial domain, since *everything* will then be a set : but that comes much later.

In set theory, there is really only one fundamental notion:

The ability to regard any collection of objects as a single entity (i.e. a set).

It is by asking ourselves what may and what may not *determine* "a collection" that we shall arrive at the axioms of set theory. For the present, we regard the two words "set" and "collection (of objects)" as synonymous, and understood.

If a is an object and x is a set, we write

$$a \in x$$

to mean that a is an *element* of (or *member* of) x, and

$$a \notin x$$

to mean that a is not an element of x.

In set theory, perhaps more than in any other branch of mathematics, it is vital to set up a collection of symbolic abbreviations for various logical concepts. Because the basic assumptions of set theory are absolutely minimal, assertions about sets tend to be logically complex, and a good system of abbreviations helps to make complex statements readable. For instance, the symbol '$\in$' has already been introduced to abbreviate the phrase 'is an element of'. We also make considerable use of the following (standard) logical symbols:

$\rightarrow$ abbreviates 'implies' ;

$\leftrightarrow$ abbreviates 'if and only if' ;

$\neg$ abbreviates 'not' ;

$\wedge$ abbreviates 'and' ;

$\vee$ abbreviates 'or' ;

$\forall$ abbreviates 'for all' ;

$\exists$ abbreviates 'there exists ... such that'.

Note that in the case of 'or', we adopt the usual, mathematical interpretation, whereby $\phi \vee \psi$ means that either ϕ is true or ψ is true, or else both ϕ and ψ are true (where ϕ, ψ denote any assertions in any language, here).

The above logical notions are not totally independent, of course. For any statements ϕ, ψ, we have, for instance :

$\phi \leftrightarrow \psi$ is the same as $(\phi \rightarrow \psi) \wedge (\psi \rightarrow \phi)$;

$\phi \rightarrow \psi$ is the same as $(\neg\phi) \vee \psi$;

$\phi \vee \psi$ is the same as $\neg((\neg\phi) \wedge (\neg\psi))$;

$\exists x \phi$ is the same as $\neg((\forall x)(\neg\phi))$;

where the phrase 'is the same as' means that the two expressions are logically equivalent.

Exercise 1: *Let $\phi \dot{\vee} \psi$ mean that exactly one of ϕ, ψ is true. Express $\phi \dot{\vee} \psi$ in terms of the symbols introduced above.*

Let us return now to the notion of a set. Since a set is the same as a collection of objects, a set will be uniquely determined once we know what its elements are. In symbols, this fact can be expressed as follows :

$$x = y \leftrightarrow \forall a[(a \in x) \leftrightarrow (a \in y)].$$

(This principle will, in fact, form one of our axioms of set theory : *the Axiom of Extensionality*.)

If x, y are sets, we say x is a *subset* of y iff every element of x is an element of y, and write

$$x \subseteq y$$

in this case. In symbols, this definition reads[†]

$$(x \subseteq y) \leftrightarrow \forall a[(a \in x) \rightarrow (a \in y)].$$

We write

$$x \subset y$$

in case x is a subset of y and x is not equal to y ; thus :

$$(x \subset y) \leftrightarrow (x \subseteq y) \wedge (x \neq y)$$

(where, as usual, we write $x \neq y$ instead of $\neg(x = y)$, just as we did earlier with $\in$). Clearly we have

$$(x = y) \leftrightarrow [(x \subseteq y) \wedge (y \subseteq x)].$$

[†]The reader should attain the facility of "reading" symbolic expressions such as this as soon as possible. In more complex situations the symbolic form is the only intelligible one.

Exercise 2: *Check the above assertion by replacing the subset symbol by its definition given above, and reducing the resulting formula logically to the axiom of extensionality. Is the above statement an equivalent formulation of the axiom of extensionality?*

2. Operations on Sets

There are a number of simple operations which can be performed on sets, forming new sets from given sets. We consider below the most common of these.

If x and y are sets, the *union* of x and y is the set consisting of the members of x together with the members of y, and is denoted by

$$x \cup y.$$

Thus, in symbols, we have

$$(z = x \cup y) \leftrightarrow \forall a[(a \in z) \leftrightarrow (a \in x \lor a \in y)].$$

In the above, in order to avoid proliferation of brackets, we have adopted the convention that the symbol $\in$ predominates over logical symbols. This convention, and a similar one for =, will be adhered to. An alternative way of denoting the above definition is :

$$(a \in x \cup y) \leftrightarrow (a \in x \lor a \in y).$$

Using this last formulation, it is easy to show that the union operation on sets is both commutative and associative : thus

$$x \cup y = y \cup x$$

$$x \cup (y \cup z) = (x \cup y) \cup z.$$

(The beginner should check this, and any similar assertions we make in this chapter.)

The *intersection* of sets x and y is the set consisting of those objects which are members of both x and y, and is denoted by

$$x \cap y.$$

Thus :

$$(a \in x \cap y) \leftrightarrow (a \in x \wedge a \in y).$$

The intersection operation is also commutative and associative.

The *difference* of sets x and y is the set consisting of those elements of x which are not elements of y, and is denoted by

$$x - y.$$

Thus :

$$(a \in x - y) \leftrightarrow (a \in x \wedge a \notin y).$$

Care should be exercised with the difference operation at first. Notice that x - y is *always* defined and is *always* a subset of x, regardless of whether y is a subset of x or not.

<u>Exercise 1</u>: *Prove the following assertions directly from the definitions.*†

(a) $x \cup x = x$; $x \cap x = x$;

(b) $x \subseteq x \cup y$; $x \cap y \subseteq x$;

(c) $[(x \subseteq z) \wedge (y \subseteq z)] \rightarrow [x \cup y \subseteq z]$;

(d) $[(z \subseteq x) \wedge (z \subseteq y)] \rightarrow [z \subseteq x \cap y]$;

---

†The drawing of 'Venn diagrams' is forbidden. For the reader who is familiar with these ideas, which probably includes you, this is an exercise in the manipulation of logical concepts.

(e) $x \cup (y \cap z) = (x \cup y) \cap (x \cup z)$;

(f) $x \cap (y \cup z) = (x \cap y) \cup (x \cap z)$;

(g) $(x \subseteq y) \leftrightarrow (x \cap y = x) \leftrightarrow (x \cup y = y)$.

Exercise 2: *Let x, y be subsets of a set z. Prove the following assertions :*

(a) $z - (z - x) = x$;

(b) $(x \subseteq y) \leftrightarrow [(z - y) \subseteq (z - x)]$;

(c) $x \cup (z - x) = z$;

(d) $z - (x \cup y) = (z - x) \cap (z - y)$;

(e) $z - (x \cap y) = (z - x) \cup (z - y)$.

Exercise 3: *Prove that for any sets x, y,*

$$x - y = x - (x \cap y).$$

In set theory, it is convenient to regard the collection of no objects as a set, the *empty set* or *null set*. This set is usually denoted by the Scandinavian letter $\emptyset$.

Exercise 4: *Prove (from the axiom of extensionality) that there is only one empty set. (This requires a sound mastery of the elementary logical concepts introduced earlier.)*

Two sets x and y are said to be *disjoint* if they have no member in common; in symbols

$$x \cap y = \emptyset.$$

Exercise 5: *Prove the following:*

(a) $x - \emptyset = x$;

(b) $x - x = \emptyset$;

(c) $x \cap (y - x) = \emptyset$

(d) $\emptyset \subseteq x$.

3. Notation for Sets

Suppose we wish to provide an accurate description of a set x. How can we do this? Well, if the set concerned is finite, we can enumerate its members. If x consists of the objects $a_1, \ldots, a_n$, we denote x by the symbol

$$\{a_1, \ldots, a_n\}.$$

Thus, the statement

$$x = \{a_1, \ldots, a_n\}$$

should be read as "x is the set whose elements are $a_1, \ldots, a_n$". For example, the *singleton* a is the set

$$\{a\} ,$$

and the *doubleton* a, b is the set

$$\{a,b\} .$$

But in the case of infinite sets, this type of notation is no longer possible. Of course, one can establish conventions, whereby symbols such as

$$\{a_1, a_2, a_3, \ldots\}$$

have an (obvious) meaning, but this does not get us very far, and is certainly not very precise (in a strict sense). Providing the set concerned is defined by some *property*, however, there is a precise notation already to hand. If x is the set of all those a for which P(a) holds, where P is some property, then we may write

$$x = \{a \mid P(a)\}.$$

Thus, for example, the set of all real numbers may be denoted by

$$\{a \mid a \text{ is a real number}\}.$$

Exercise 1: *Prove the following equalities :*

(a) $x \cup y = \{a \mid a \in x \vee a \in y\}$;

(b) $x \cap y = \{a \mid a \in x \wedge a \in y\}$;

(c) $x - y = \{a \mid a \in x \wedge a \notin y\}$.

4. Sets of Sets

So far, we have been tacitly distinguishing between sets and objects. Admittedly, we did not restrict in any way the choice of initial objects - they could themselves be sets - but we distinguished between these initial objects and the sets of those objects which we could form. But, as we said at the beginning, the main idea in set theory is that any collection of objects can be regarded as a single entity (i.e. a *set*). Thus we are entitled to build sets out of entities which are themselves sets. Commencing with some given domain of objects then, we can first build sets of objects, then sets of sets of objects, and so on. Indeed, we can make more complicated sets, some of whose elements are basic objects, and some of which are sets of basic objects, etc. In this manner one can *define* the *ordered pair* of two objects a, b by:

$$(a,b) = \{ \{a\}, \{a,b\} \}.$$

Thus (a,b) is a set : it is a set of sets of objects.

<u>Exercise 2</u>: *Show that the above definition does define an ordered pair operation; i.e. prove that for any* a, b, a', b',

$$(a,b) = (a',b') \leftrightarrow (a = a' \wedge b = b').$$

(*Don't forget the case* $a = b$!)

The n-tuple $(a_1, \ldots, a_n)$ may now be defined iteratively, thus :

$$(a_1, \ldots, a_n) = ((a_1, \ldots, a_{n-1}), a_n).$$

It is clear that $(a_1, \ldots, a_n) = (a_1', \ldots, a_n')$ iff $a_1 = a_1' \wedge \ldots \wedge a_n = a_n'$. Of course, it is not important how one defines an ordered pair operation. What counts is its behaviour. Thus, the property described in Exercise 2 above is the only requirement we have of an ordered pair. So in *naive* set theory, one could just take (a,b) as a basic, undefined operation from pairs of objects to objects. But it is perhaps more convenient to define it as a certain set, if this is possible (which, as we have seen, it is), and anyway, when we come to axiomatic set theory this will be necessary. There are other definitions, but the one given is the most common, and it is the one we shall adopt throughout this book.

Another way of defining "new" sets from "old" ones is via the *power set operation*. If x is any set, the collection of all subsets of x is a well-defined collection of objects (= sets, in this case), and hence may be regarded as an entity (= set) in its own right† : the *power set* of x, $\mathcal{P}(x)$. Thus

$$\mathcal{P}(x) = \{y \mid y \subseteq x\}.$$

Suppose now that x is a set of sets of objects. The *union* of x is the set of all elements of all elements of x, and is denoted by $\cup x$. Thus

$$\cup x = \{a \mid \exists y(y \in x \wedge a \in y)\}.$$

---

†For the benefit of the experienced reader, let us point out that in *naive* set theory, almost anything goes.

Extending our logical notation by writing

$$(\exists y \in x)$$

to mean "there exists a y in x such that", this may also be written

$$\cup x = \{a \mid (\exists y \in x)(a \in y)\}.$$

The *intersection* of x is the set of all objects which are elements of all elements of x, and is denoted by $\cap x$. Thus

$$\cap x = \{a \mid \forall y(y \in x \rightarrow a \in y)\}.$$

Or, more succinctly,

$$\cap x = \{a \mid (\forall y \in x)(a \in y)\},$$

where $(\forall y \in x)$ means "for all y in x".

If $x = \{y_i \mid i \in I\}$ (so I is some indexing set for the elements of x), we often write

$$\bigcup_{i \in I} y_i$$

for $\cup x$, and

$$\bigcap_{i \in I} y_i$$

for $\cap x$.

This ties in with our earlier notation to some extent, since we clearly have, for any sets x, y

$$x \cup y = \cup\{x,y\} ;$$

$$x \cap y = \cap\{x,y\} .$$

Exercise 3: (a) *Prove the above equalities.*

(b) *What are* $\cup\{x\}$ *and* $\cap\{x\}$?

(c) *What are* $\cup\emptyset$ *and* $\cap\emptyset$?

Exercise 4: *Prove that if* $\{x_i \mid i \in I\}$ *is a family of sets, then:*

(a) $\bigcup_{i\in I} x_i = \{a \mid (\exists i \in I)(a \in x_i)\}$;

(b) $\bigcap_{i\in I} x_i = \{a \mid (\forall i \in I)(a \in x_i)\}$.

Exercise 5: *Prove the following :*

(a) $(\forall i \in I)(x_i \subseteq y) \rightarrow (\bigcup_{i\in I} x_i \subseteq y)$;

(b) $(\forall i \in I)(y \subseteq x_i) \rightarrow (y \subseteq \bigcap_{i\in I} x_i)$;

(c) $\bigcup_{i\in I} (x_i \cup y_i) = (\bigcup_{i\in I} x_i) \cup (\bigcup_{i\in I} y_i)$;

(d) $\bigcap_{i\in I} (x_i \cap y_i) = (\bigcap_{i\in I} x_i) \cap (\bigcap_{i\in I} y_i)$;

(e) $\bigcup_{i\in I} (x_i \cap y) = (\bigcup_{i\in I} x_i) \cap y$;

(f) $\bigcap_{i\in I} (x_i \cup y) = (\bigcap_{i\in I} x_i) \cup y$.

Exercise 6: *Let* $\{x_i \mid i \in I\}$ *be a family of subsets of* z. *Prove:*

(a) $z - \bigcup_{i\in I} x_i = \bigcap_{i\in I} (z - x_i)$;

(b) $z - \bigcap_{i\in I} x_i = \bigcup_{i\in I} (z - x_i)$.

5. Relations

If x, y are sets, the *cartesian product* of x and y is the set

$$x \times y = \{(a,b) \mid a \in x \wedge b \in y\}.$$

More generally, if $x_1, \ldots, x_n$ are sets, we define their *cartesian product* by

$$x_1 \times \ldots \times x_n = \{(a_1,\ldots,a_n) \mid a_1 \in x_1 \wedge \ldots \wedge a_n \in x_n\}.$$

A *unary relation* on a set x is a subset of x. An *n-ary relation* on x, for $n > 1$, is a subset of the n-fold cartesian product $x \times \ldots \times x$. Notice that an n-ary relation on x is a unary relation on the n-fold product $x \times \ldots \times x$. These formal definitions clearly provide a concrete realisation in terms of set theory of the intuitive concept of a relation. As is often the case in set theory, having once seen how a concept may be defined set theoretically, we revert at once to the more familiar notations. Thus, if P is some property on the pairs of x, for example, we often speak of "the binary relation P on x", though strictly speaking, the relation concerned is the set

$$\{(a,b) \mid a \in x \ \wedge \ b \in x \ \wedge \ P(x,y)\}.$$

Also common is the tacit *identification* of P with the relation it defines, so that $P(a,b)$ and $(a,b) \in P$ mean the same. Similarly, going in the opposite direction, if R is some binary relation on x, say, we will write $R(a,b)$ instead of $(a,b) \in R$. Indeed, in the specific case of binary relations, we sometimes go even further, writing aRb instead of $R(a,b)$. In the case of ordering relations, this notation is, of course, well known : we rarely write $<(a,b)$ or $(a,b) \in <$, though both could be said to be more accurate than the more common $a < b$ from a set theoretic point of view.

The binary relations play a particularly important role in set theory (and, indeed, in mathematics as a whole). So much so, that we shall devote the rest of this section to a study of binary relations.

There are several properties which a binary relation may fulfill. Let R denote any binary relation on a set x. We say :-

R is *reflexive* if $(\forall a \in x)(aRa)$;

R is *symmetric* if $(\forall a,b \in x)(aRb \rightarrow bRa)$;

R is *antisymmetric* if $(\forall a,b \in x)[(aRb \wedge a \neq b) \rightarrow (\neg bRa)]$;

R is *connected* if $(\forall a,b \in x)[(a \neq b) \rightarrow (aRb \vee bRa)]$;

R is *transitive* if $(\forall a,b,c \in x)[(aRb \wedge bRc) \rightarrow (aRc)]$.

(Notice the obvious use of the repeated quantifier in the above, writing, for example, $(\forall a,b \in x)$ instead of the more cumbersome $(\forall a \in x)(\forall b \in x)$.)

<u>Exercise 1</u>: *Which of the above properties does the membership relation,* $\in$*, on a set satisfy?*

A binary relation on a set is an *equivalence relation* just in case it is reflexive, symmetric, and transitive. If R is an equivalence relation on a set x, the *equivalence class* of an element a of x under the equivalence relation R is the set

$$[a] = [a]_R = \{b \in x \mid aRb\}.$$

<u>Exercise 2</u>: *Let* R *be an equivalence relation on a set* x. *Then* R *partitions* x *into a collection of disjoint equivalence classes.*

Examples of equivalence relations pervade the whole of contemporary pure mathematics. So too do examples of our next concept, that of an *ordering relation*.

A *partial ordering* of a set x is a binary relation on x which is reflexive, antisymmetric, and transitive. Usually (but not always) partial orderings are denoted by the symbol $\leq$. A *partially ordered set* (or *poset*) consists of a set, x, together with a partial ordering, $\leq$, of that set. More formally, we define the ordered pair $(x, \leq)$ *to be the poset* here. Let $(x, \leq)$ be a poset, and let $y \subseteq x$. An element a of y is a *minimal* element of y iff there is no b in y such that $b < a$ (where, as usual, we write $b < a$ to denote $b \leq a \wedge b \neq a$). A poset $(x, \leq)$ is said to be *well-founded* if every non-empty subset of x has a minimal element. (Equivalently, we often say that the ordering relation $\leq$ is *well-founded*.)

5.1 Lemma

Let $(x, \leq)$ be a poset. $(x, \leq)$ is well-founded iff there is no sequence $\{a_n\}$ of elements of x such that $a_{n+1} < a_n$ for all n (i.e. no sequence $\{a_n\}$ such that $a_1 > a_2 > a_3 > \ldots$).

Proof : Suppose $(x, \leq)$ is not well-founded. Let $y \subseteq x$ have no minimal element. Let $a_1 \in y$. Since a_1 is not minimal in y, we can find $a_2 \in y$, $a_2 < a_1$. Again, a_2 is not minimal in y, so we can find $a_3 \in y$, $a_3 < a_2$. Proceeding inductively, we obtain a sequence $a_1 > a_2 > a_3 > \ldots$.

Now suppose there is a sequence $a_1 > a_2 > a_3 > \ldots$. Let $y = \{a_1, a_2, a_3, \ldots\}$. Clearly, y has no minimal member. □

The subset relation, $\subseteq$, on the subsets of a set x clearly constitutes a partial ordering of the power set of x, $\mathcal{P}(x)$. Indeed, the subset relation on any collection of sets is a partial ordering of that collection. And, in fact, this is, up to isomorphism, the only type of partial ordering there is:

5.2 Theorem

Let $(x, \leq)$ be a poset. Then there is a set y of subsets of x such that $(x, \leq) \cong (y, \subseteq)$.

Proof : For each $a \in x$, let $z_a = \{b \in x \mid b \leq a\}$, and let $y = \{z_a \mid a \in x\}$. Define $\pi : x \to y$ by $\pi(a) = z_a$. Clearly, π is a bijection. Moreover, $a_1 \leq a_2 \leftrightarrow z_{a_1} \subseteq z_{a_2}$, so π is an isomorphism between $(x, \leq)$ and $(y, \subseteq)$. □

A *total ordering* (or *linear ordering*) of a set x is a connected partial ordering of x. A *totally ordered set* (or *toset*) is a pair $(x, \leq)$ such that $\leq$ is a total ordering of the set x.

A *well-ordering* of a set x is a well-founded total ordering of x. A *well-ordered set* (or *woset*) is a pair $(x, \leq)$ such that $\leq$ is a well-ordering of x. The concept of a well-ordering is central in set theory, but the reader must wait for a while before he discovers this fact from our account, as we leave these notions for the time being.

6. Functions

We all know, more or less, what a function is. Indeed, in §5 we have already made use of functions in stating and proving 5.2. But there we followed the usual mathematical practice of using the function concept without worrying too much about what a function really *is*.

Let R be an (n+1)-ary relation on a set x. The *domain* of R is the set

$$\mathrm{dom}(R) = \{ a \mid \exists b [(a,b) \in R] \}.$$

The *range* of R is the set

$$\mathrm{ran}(R) = \{ b \mid \exists a [(a,b) \in R] \}.$$

Notice that we have not used the loose notation R(a,b) here. This is to emphasise the way the definition of the ordered (n+1)-tuple is used in these definitions. In case $n = 1$ and R is a binary relation, it is clear what is meant : dom(R) is the set of first components of R, ran(R) the set of second components. But what if $n > 1$? In this case, any member of R will be an (n+1)-tuple. But what is an (n+1)-tuple? Any (n+1)-tuple, c, has the form (a,b) where a is an n-tuple and b is an object in x. In other words, even if $n > 1$, the elements of R will still be ordered pairs, only now the domain of R will consist not of elements of x but of elements of the n-fold product $x \times \ldots \times x$. Although the notions of domain and range for an arbitrary relation are quite common in more advanced parts of set theory, chances are that the reader is not used to these concepts. But when we define the notion of a function, below, as a special sort of relation, the reader will see at once that the above definitions coincide with what one usually means by the "domain" and "range" of a function.

An n-ary *function* on a set x is an (n+1)-ary relation, R, on x such that for every $a \in dom(R)$ there is *exactly one* $b \in ran(R)$ such that $(a,b) \in R$. As usual, if R is an n-ary function on x and $a_1, \ldots, a_n, b \in x$, we write

$$R(a_1, \ldots, a_n) = b$$

instead of

$$(a_1, \ldots, a_n, b) \in R .$$

Exercise 1: *Comment on the assertion that a set-theorist is a person to whom all functions are unary. (This is a serious exercise, and concerns a subtle point which often causes problems for the beginner.)*

If f is a function and $dom(f) = x$, $ran(f) \subseteq y$, we write $f : x \to y$. Notice that if $f : x \to y$, then $f \subseteq x \times y$.

A *constant function* on a set x is a function on x of the form $f = \{ (a,k) \mid a \in dom(f) \}$, where k is a fixed member of x. The *identity function* on x is the unary function defined by

$$id_x = \{ (a,a) \mid a \in x \}.$$

If $f : x \to y$ and $g : y \to z$, we define $g \circ f : x \to z$ by

$$g \circ f(a) = g(f(a)).$$

Exercise 2: *Express* $g \circ f$ *as defined above as a set of ordered pairs.*

Let $f : x \to y$. If $u \subseteq x$, we set

$$f[u] = \{ f(a) \mid a \in u \},$$

the *image* of u under f. And if $v \subseteq y$, we set

$$f^{-1}[v] = \{a \in x \mid f(a) \in v\},$$

the *preimage* of v under f.

<u>Exercise 3</u>: *Let* $f : x \to y$, *and let* $v_i \subseteq y$ *for* $i \in I$. *Prove that:*

(a) $f^{-1}[\bigcup_{i\in I} v_i] = \bigcup_{i\in I} f^{-1}[v_i]$;

(b) $f^{-1}[\bigcap_{i\in I} v_i] = \bigcap_{i\in I} f^{-1}[v_i]$;

(c) $f^{-1}[v_i - v_j] = f^{-1}[v_i] - f^{-1}[v_j]$.

If $f : x \to y$ and $u \subseteq x$, we define

$$f \restriction u = \{ (a,f(a)) \mid a \in u \},$$

the *restriction* of f to u. Notice that $f \restriction u$ is a *function*, with domain u.

<u>Exercise 4</u>: (a) *Prove that if* $f : x \to y$ *and* $u \subseteq x$, *then*

$$f[u] = \mathrm{ran}(f \restriction u).$$

(b) *Prove that if* $f : x \to y$ *and* $u \subseteq x$, *then*

$$f \restriction u = f \cap (u \times \mathrm{ran}(f)).$$

Let $f : x \to y$. We say f is *injective* (or *one-one*) iff

$$a \neq b \to f(a) \neq f(b).$$

We say f is *surjective* (or *onto*) (relative to the stated set y) iff

$$f[x] = y.$$

And we say f is *bijective* iff it is both injective and surjective. In this last case we often write $f : x \leftrightarrow y$.

If $f : x \to y$ is bijective, then f has a unique *inverse function*, f^{-1}, defined by

$$f^{-1} = \{ (b,a) \mid (a,b) \in f \}.$$

Thus, $f^{-1} : y \to x$ and $f^{-1} \circ f = id_x$ and $f \circ f^{-1} = id_y$. It should be noticed that whenever $f : x \to y$ and $v \subseteq y$, the set $f^{-1}[v]$ is defined, regardless of whether f is bijective (and hence has an inverse function) or not. If, in fact, f is bijective, and f^{-1} exists, then the two possible interpretations of $f^{-1}[v]$ clearly coincide, so the choice of notation should cause no problems.

Having defined the notion of a function now, we may give a very general definition of a "cartesian product" of an arbitrary (not necessarily finite) family of sets.

Let x_i, $i \in I$, be a family of sets. The *cartesian product* of the family $\{x_i \mid i \in I\}$ is the set

$$\bigtimes_{i\in I} x_i = \{f \mid (f:I \to \bigcup_{i\in I} x_i) \wedge (\forall i \in I)(f(i) \in x_i)\}.$$

If $x_i = x$ for all $i \in I$, we write x^I instead of $\bigtimes_{i\in I} x_i$. Now, in case I is finite, this provides us with a second definition of "cartesian product", quite different from the first. However, though different, the two types of finite cartesian product are clearly closely related, and either definition of product may be used. In general, one uses the original definition for finite products, the above definition for infinite products (or arbitrary products). This means that the notation will indicate the usage.

<u>Exercise 5</u>: *What <u>set</u> is the cartesian product* x^I?

<u>Exercise 6</u>: *The ordered pair operation* (a,b) *defines a binary function on sets. The inverse functions to the function are defined as follows:*

if $w = (a,b)$, *then* $(w)_0 = a$ *and* $(w)_1 = b$.

Prove that if w *is an ordered pair, then :*

(a) $(w)_0 = \cup\cap w$;

(b) $(w)_1 = \begin{cases} \cup[\cup w - \cap w], & \text{if } \cup w \neq \cap w \\ \cup\cup w, & \text{if } \cup w = \cap w . \end{cases}$

(To avoid unnecessary complication, we have not bothered to specify the set on which the above functions are defined. This is, of course, common mathematical practice, when one is only interested in the behaviour of the functions concerned.)

7. Well-Orderings and Ordinals

We promised earlier that well-orderings would return, and here they come. Firstly, let us try to indicate why they play such an important role in set theory. Well, everyone is familiar with the principle of mathematical induction in establishing results about the positive integers. Indeed, this method is not restricted to proving results about the integers, but will work for any set which may be enumerated as a sequence $\{a_n\}$ indexed by the integers. Readers familiar with the Peano axioms for the natural numbers will already be aware of the fact that what makes the induction method work is the fact that the positive integers are well-ordered. There is, after all, no real possibility of ever proving, *case by case*, that P(n) holds for every n. But since the positive integers are well-ordered, if P(n) is ever to fail, it will fail at a least n, and then we would have P(n-1) true but P(n) false, and it is precisely this situation which we exclude in our "induction proof". How nice it would be if we could extend this powerful method of proof to cover transfinite sets, not enumerable as an integer-indexed sequence. Well, a natural place to start would be by considering an extension of the positive integers into the transfinite, to obtain a system of numbers suitable for enumerating any set, however large. We do this by adopting the same method (more or less) that a small child uses when learning the number concept. The child first learns to count collections (by enumerating them in a linear way), and then, after repeating this process many times, abstracts from it the concept of 'number'. This is just what we will do, only in a more formal manner. Of course, since we are going to allow infinite collections, we shall not be doing any actual "counting", but the concept of a well-ordering will provide the mathematical counterpart to this.

Recall that a well-ordering of a set x is a total ordering of x which is well-founded. Now, according to our previous definition, a partial ordering of a set x is well-founded iff every non-empty subset of x has a minimal element (i.e. an element having no predecessor in the subset). But in the case of total orderings an element of a subset y of x will be minimal iff it is the (unique) smallest member of y. Thus a well-ordering of a set x is a total ordering of x such that every non-empty subset of x has a smallest member. This enables us to prove:

7.1 Theorem (Principle of Proof by Induction on a Well-Ordering)

Let $(X, \leq)$ be a woset. Let E be a subset of X such that:

(i) the smallest element of X is a member of E;

(ii) for any $x \in X$, if $y < x \rightarrow y \in E$, then $x \in E$.

Then $E = X$.

Proof: Suppose $E \neq X$. Let x be the smallest member of the non-empty set $X - E$. Then x is not the smallest member of X, by (i). But by choice of x, $y < x \rightarrow y \in E$. Hence by (ii), $x \in E$, a contradiction. □

Notice the notation adopted above. We used capital letters to denote sets, lower case letters to denote their elements. This is a very common notation, which we shall often adopt. (It is really only helpful in simple situations. Once there are sets of sets floating about it becomes rather confusing.)

Now, 7.1 allows us to prove results by induction on a well-founded set, but it does not provide us with a system of transfinite numbers for "counting". For that we need to isolate just what it is that all wosets have in common. So we commence by comparing wosets.

Let $(X, \leq)$, $(X', \leq')$ be wosets. A function $f : X \to X'$ is an *order isomorphism* iff f is bijective and

$$x < y \to f(x) <' f(y).$$

We write $f : X \cong X'$ in this case. (As usual, we adopt the principle of writing X in place of $(X, \leq)$, etc., it being clear from the context that X is a set with a well-ordering here.)

7.2 Theorem

Let $(X, \leq)$ be a woset, $Y \subseteq X$, $f : X \cong Y$. Then for all $x \in X$, $x \leq f(x)$.

Proof: Let $E = \{x \in X \mid f(x) < x\}$. We must prove that $E = \emptyset$. Suppose otherwise. Then E has a smallest member, x_o. Since $x_o \in E$, $f(x_o) < x_o$. Let $x_1 = f(x_o)$. Since $x_1 < x_o$, applying f gives $f(x_1) < f(x_o)$. Thus $f(x_1) < x_1$. Thus $x_1 \in E$. But $x_1 < x_o$, so this contradicts the choice of x_o as the least member of E. □

7.3 Theorem

Let $(X, \leq)$, $(X', \leq')$ be wosets. If $(X, \leq) \cong (X', \leq')$, there is exactly one order isomorphism $f : X \cong X'$.

Proof: Let $f : X \cong X'$, $g : X \cong X'$. Set $h = f^{-1} \circ g$. It is easily seen that $h : X \cong X$. So, by 7.2, $x \leq h(x)$ for all x in X. So, applying f, we see that for any x in X, $f(x) \leq f(h(x)) = g(x)$. Similarly, $g(x) \leq f(x)$ for any x in X. Thus $f = g$. □

It should be noticed that the above result does not hold for any tosets, but that well-ordering is essential. For example, let $\mathbb{Z}$ be the set of all integers, $\leq$ the usual ordering on $\mathbb{Z}$. For any m, the mapping $f_m : \mathbb{Z} \to \mathbb{Z}$ defined by $f_m(n) = n+m$ is an order isomorphism, and $m \neq m'$ implies $f_m \neq f_{m'}$. Notice also that if $m < 0$, then $f_m(n) < n$ for all n, so this example also shows that 7.2 also requires well-ordering.

Let $(X, \leq)$ be a woset, $a \in X$. The *segment* X_a of X determined by a is the set

$$X_a = \{x \in X \mid x < a\} .$$

7.4 <u>Theorem</u>

Let $(X, \leq)$ be a woset. There is no isomorphism of X onto a segment of X.

Proof: Suppose $f : X \cong X_a$. By 7.2, $x \leq f(x)$ for all x in X. In particular, therefore, $a \leq f(a)$. But $\text{ran}(f) = X_a$, so $f(a) \in X_a$, so $f(a) < a$. Contradiction.□

Notice that well-ordering is required for 7.4. For example, let $\mathbb{Z}^-$ denote the non-positive integers, and define $f : \mathbb{Z}^- \cong \mathbb{Z}^-_0$ by $f(n) = n-1$.

7.5 <u>Theorem</u>

Let $(X, \leq)$ be a woset, $A = \{X_a \mid a \in X\}$. Then

$$(X, \leq) \cong (A, \subseteq) .$$

Proof : Define $f : X \cong A$ by $f(a) = X_a$. □

A woset $(X, \leq)$ such that $X_a = a$ for all a in X will be called an <u>ordinal</u>. (We are not saying anything about the existence of such sets at the moment.)

<u>Exercise 1</u> : *Suppose* $(X, \le)$ *is an ordinal. What must the first member of* X *be? Well, if* x_o *is the first member of* X, *then* $X_{x_o} = \emptyset$, *so as* $(X, \le)$ *is an ordinal,* $x_o = X_{x_o} = \emptyset$. *Now what must the second member,* x_1, *of* X *be? In general, what is the* n-th *member of* X? *What can you guess about both the existence and uniqueness of ordinals?*

Let $(X, \le)$ be an ordinal. Then, for x, y in X, we have

$$x < y \quad \text{iff} \quad X_x \subset X_y \quad \text{iff} \quad x \subset y.$$

(The first equivalence here holds for any woset, the second because $X_x = x$ and $X_y = y$ for an ordinal.) Hence the ordering of X is the subset relation. In other words, when we specify an ordinal, we do not have to say what the ordering is; it must be the subset relation.

7.6 <u>Theorem</u>

Let X be an ordinal. If $a \in X$, then X_a is an ordinal.

Proof : Let $b \in X_a$. Then

$$(X_a)_b = \{x \in X_a \mid x < b\} = \{x \in X \mid x < a \wedge x < b\}$$

$$= \{x \in X \mid x < b\} = X_b = b. \qquad \square$$

7.7 <u>Theorem</u>

Let X be an ordinal. Let $Y \subset X$. If Y is an ordinal, then $Y = X_a$ for some $a \in X$.

Proof: Let a be the smallest element of X - Y. Thus $X_a \subseteq Y$. Now let $b \in Y$. Then $Y_b = b = X_b$, so if $a < b$, then $a \in X_b$, so $a \in Y_b$, so $a \in Y$, which is not the case. Hence $b \le a$. But $b \ne a$, since $b \in Y$. Hence $b < a$. Thus $b \in X_a$. This proves that $Y \subseteq X_a$. Hence $Y = X_a$. $\square$

7.8 Theorem

If X, Y are ordinals, then $X \cap Y$ is an ordinal.

Proof : Let $a \in X \cap Y$. Then $X_a = a = Y_a$; i.e.

$$\{x \in X \mid x < a\} = a = \{y \in Y \mid y < a\}.$$

Hence
$$a = \{z \in X \cap Y \mid z < a\} = (X \cap Y)_a. \qquad \square$$

7.9 Theorem

Let X, Y be ordinals. If $X \neq Y$, then one is a segment of the other.

Proof: If $X \subset Y$ or $Y \subset X$, we are done by 7.7. So suppose otherwise. Thus $X \cap Y \subset X$ and $X \cap Y \subset Y$. Now, by 7.8, $X \cap Y$ is an ordinal, so by 7.7, $X \cap Y = X_a$ for some $a \in X$ and $X \cap Y = Y_b$ for some $b \in Y$. Then, '

$$a = X_a = X \cap Y = Y_b = b.$$

But $a \in X$, $b \in Y$. Thus $a = b \in X \cap Y$. But $X \cap Y = X_a$, so $x \in X \cap Y \rightarrow x < a$, and we have a contradiction. $\square$

7.10 Theorem

If X, Y are isomorphic ordinals, then $X = Y$.

Proof: Let $f : X \cong Y$. We prove that $f = id_X$. Set

$$E = \{x \in X \mid f(x) \neq x\}.$$

We must prove that $E = \emptyset$. Suppose otherwise, and let a be the smallest member of E. Then $x < a \rightarrow f(x) = x$, so $X_a = Y_{f(a)}$. But then $a = X_a = Y_{f(a)} = f(a)$, contrary to $a \in E$. $\square$

7.11 Theorem

Let $(X, \leq)$ be a woset such that for each $a \in X$, X_a is isomorphic to an ordinal. Then X is isomorphic to an ordinal.

Proof : For each $a \in X$, let $g_a : X_a \cong Z(a)$ be an isomorphism of X_a onto an ordinal $Z(a)$. By 7.10 and 7.3, both $Z(a)$ and g_a are unique. Hence this defines a function Z on X. Let W be its range[†], i.e. $W = \{Z(a) \mid a \in X\}$. Define $f : X \to W$ by

$$f(a) = Z(a).$$

Claim : If $x, y \in X$, then $x < y \to Z(x) \subset Z(y)$.

Proof of claim : Let $x, y \in X$, $x < y$. Then

(1) $g_x : X_x \cong Z(x)$.

Also, since $X_x = \{z \in X \mid z < x\} = \{z \in X \mid z < y \wedge z < x\} = \{z \in X_y \mid z < x\} = (X_y)_x$, we have

(2) $(g_y \upharpoonright X_x) : X_x \cong (Z(y))_{g_y(x)}$.

Now, Z(y) is an ordinal, so by 7.6, $(Z(y))_{g_y(x)}$ is an ordinal. But by (1) and (2), $Z(x) \cong (Z(y))_{g_y(x)}$. Hence by 7.10,

(3) $Z(x) = (Z(y))_{g_y(x)}$.

Thus, in particular, $Z(x) \subset Z(y)$. The claim is proved.

[†]When we come to describe the axioms of set theory, the reader will be able to see that what we are actually doing here is applying the Axiom of Replacement. So this step is, in fact, one of the deeper steps in our present development. If the reader finds this footnote confusing, it just demonstrates what a natural principle the Axiom of Replacement is.

By the claim, f is a bijection of X onto W. Also by the claim, f is an order isomorphism of X onto the poset $(W, \subseteq)$. Thus, in particular, W is well-ordered by $\subseteq$. We finish the proof by showing that W is an ordinal. Let $y \in X$. Since $Z(y)$ is an ordinal, we have

$$x < y \rightarrow (Z(y))_{g_y(x)} = g_y(x) .$$

So by (3),

$$(4) \quad x < y \rightarrow Z(x) = g_y(x).$$

Hence,

$$\begin{aligned} W_{Z(y)} &= \{ Z(x) \mid Z(x) \subset Z(y) \} \\ &= \{ Z(x) \mid x < y \} \\ &= \{ g_y(x) \mid x < y \} \\ &= g_y[X_y] \\ &= Z(y). \end{aligned}$$

Thus, as $Z(y)$ was an arbitrary member of W (since y was an arbitrary member of X), W is an ordinal. □

Exercise 2 : *During the course of the above proof, we emphasised one point by a footnote. From the point of view of naive set theory, there was no problem : our argument was a sound mathematical argument. But when we come to axiomatise set theory we shall want to state explicitly all procedures which may be used to construct sets. Try to formulate in a precise manner the construction principle we used at the crucial part of the proof of 7.11. (The footnote may be of some assistance here.)*

7.12 <u>Theorem</u>

Every woset is isomorphic to a unique ordinal.

Proof : The uniqueness assertion follows from 7.10. We prove existence. Let $(X, \leq)$ be a woset. By 7.11 it suffices to prove that for every $a \in X$, X_a is isomorphic to an ordinal. Let

$$E = \{a \in X \mid X_a \text{ is not isomorphic to an ordinal}\}.$$

We must show that $E = \emptyset$. Suppose otherwise. Let a be the smallest element of E. Thus if $x < a$, X_x is isomorphic to an ordinal. But for $x < a$, $X_x = (X_a)_x$. Hence every segment of X_a is isomorphic to an ordinal. Hence by 7.11, X_a is isomorphic to an ordinal, contrary to $a \in E$. □

If $(X, \leq)$ is a woset, we denote by Ord(X) the unique ordinal isomorphic to X. Clearly, if X, Y are wosets, we shall have $X \cong Y$ iff $\mathrm{Ord}(X) = \mathrm{Ord}(Y)$. Since the ordinals have a certain uniqueness property (in the sense of 7.10), this means that we may use the ordinals as a yardstick for the "measurement" of the "length" of any woset : Ord(X) being the "length" of the woset X. But just how reasonable is it to take the ordinals, as defined above, as a system of numbers, which is what we are now proposing? Well, by 7.9, the ordinals are totally ordered by $\subset$. In fact 7.9 tells us more : if X, Y are ordinals, then

$$\begin{aligned} X \subset Y \quad & \text{iff} \quad X = Y_a \quad \text{for some} \quad a \in Y \\ & \text{iff} \quad X = a \quad (\text{since } Y_a = a) \\ & \text{iff} \quad X \in Y. \end{aligned}$$

Thus the ordering $\subset$ on ordinals and the ordering $\in$ on ordinals are identical. This implies also that the ordinals are well-ordered by $\subset$ (or, equivalently, by $\in$). We make use of 5.1. Suppose the ordinals were not well-ordered by $\subset$. Then we could find a sequence $\{X(n)\}$ of ordinals such that $X(1) \supset X(2) \supset X(3) \supset \ldots$ Now, for all

$n > 1$, $X(n) \subset X(1)$, so $X(n) \in X(1)$. Thus $\{X(n+1)\}$ is a decreasing (under $\subset$) sequence of members of $X(1)$. But since $X(1)$ is an ordinal, it is well-ordered by $\subset$, so we have a contradiction. It would seem, therefore, that the ordinals constitute an eminently reasonable number system, suitable for "measuring" the "length" of any woset.

It is common in contemporary set theory to reserve lower case Greek letters $\alpha, \beta, \gamma, \ldots$ to denote ordinals. (Since the ordering of an ordinal is always $\subset$, there is, of course, no need to specify this each time. But it should be remembered that an ordinal is, strictly speaking, a well-ordered set.) It is also customary to denote the order relation between ordinals by

$$\alpha < \beta$$

instead of the two equivalent forms

$$\alpha \subset \beta, \quad \alpha \in \beta,$$

though the latter is also quite common.

Now, since the ordinals will measure any woset, they will certainly measure any finite woset. But so too will the positive integers. So do we have some duplication here? Well no, because in mathematics one (almost) never bothers to *define* the integers as specific objects. As a result of our development of ordinals, we obtain, *gratis*, a neat definition of the natural numbers as specific sets : the finite ordinals.

We know that the ordinals are well-ordered by $\subset$. So what is the first ordinal? Well, if α is an ordinal, then by definition we will have

$$\alpha = \{\beta \mid \beta < \alpha\}.$$

That is, an ordinal is the set of all smaller ordinals. But in the case of the first ordinal, there is no smaller ordinal. Hence the first ordinal must be the empty set, $\emptyset$ (regarded as a well-ordered set). Let us denote this ordinal by the symbol 0. Thus, by definition, ignoring the well-ordering as usual,

$$0 = \emptyset .$$

What is the second ordinal? Well, it has to be the set of all smaller ordinals, so if we denote the second ordinal by 1, we must have

$$1 = \{0\} .$$

The third ordinal, which we denote by 2, is

$$2 = \{0, 1\} .$$

The pattern is now clear. We have

$$3 = \{0, 1, 2\} ,$$

$$4 = \{0, 1, 2, 3\} ,$$

and in general,

$$n = \{0, 1, 2, \ldots, n-1\} .$$

Notice that the ordinal n is a set with exactly n elements, making the finite ordinals ideal for "measuring" finite sets.

Notice also that if α, β are finite ordinals, then one must be a segment of the other (unless they are equal) and (hence) an element of the other.

What will be the first infinite ordinal? Clearly, it must be

$$\{0, 1, 2, \ldots, n, n+1, \ldots\} .$$

We denote this ordinal by ω. And the next? Clearly

$$\{0, 1, 2, \ldots, n, n+1, \ldots, \omega\} .$$

In general, if α is an ordinal, the next ordinal will be

$$\alpha \cup \{\alpha\} .$$

It is customary to denote the first ordinal after α by $\alpha+1$, the *successor* of α.

And if (as in the case of ω above)

$$0, 1, 2, \ldots, \omega, \omega+1, \ldots, \alpha, \alpha+1, \ldots$$

is a listing of some initial segment of the well ordered collection of ordinals having no greatest member, then the next ordinal will be

$$\{0, 1, 2, \ldots, \omega, \omega+1, \ldots, \alpha, \alpha+1, \ldots\} \ .$$

Since such an ordinal will have no greatest member, it cannot be the successor of any ordinal. Such an ordinal is called a *limit ordinal*. For example, ω is a limit ordinal.

The ordinals thus provide us with a continuation of the natural numbers into the transfinite. Further discussion of ordinals will have to be postponed until we have developed the axiomatic foundation of our set theory. But we do need the concept of a *sequence*.

A *sequence* is a function whose domain is an ordinal. If f is a sequence and $\text{dom}(f) = \alpha$, we say f is an α-*sequence*. If $f(\xi) = x_\xi$ for all $\xi < \alpha$, we often write $\langle x_\xi \mid \xi < \alpha \rangle$ in place of f. Then, for $\beta < \alpha$, $\langle x_\xi \mid \xi < \beta \rangle$ denotes $f \upharpoonright \beta$. This clearly gives a precise meaning to what we generally think of as a (transfinite, perhaps) sequence. (The "sequences" of elementary analysis are just ω-sequences, of course.)

Exercise 3. *We have already introduced the notation* $\alpha+1$ *for the next ordinal after* α. *Let us denote by* $\alpha+n$ *the* n-th *ordinal after* α, *where* n *is any natural number. Show that if* α *is any ordinal, either* α *is a limit ordinal or else there is a limit ordinal* β *and a natural number* n *such that* $\alpha = \beta+n$. *(Hint. Use 7.1)*

Problems

1. (Boolean Algebras) A *boolean algebra* is a structure consisting of a set B with a unary operation $-$ (complement) and two binary operations $\wedge$ (meet) and $\vee$ (join). The axioms to be satisfied by this structure are :

(B1) $b \vee c = c \vee b$; $b \wedge c = c \wedge b$;

(B2) $b \vee (c \vee d) = (b \vee c) \vee d$; $b \wedge (c \wedge d) = (b \wedge c) \wedge d$;

(B3) $(b \wedge c) \vee c = c$; $(b \vee c) \wedge c = c$;

(B4) $b \wedge (c \vee d) = (b \wedge c) \vee (b \wedge d)$; $b \vee (c \wedge d) = (b \vee c) \wedge (b \vee d)$;

(B5) $(b \wedge -b) \vee b = b$; $(b \vee -b) \wedge b = b$.

(A) The elements $b \wedge -b$ are all unique, and denoted by $\mathbb{0}$ (zero).

(B) The elements $b \vee - b$ are all unique, and denoted by $\mathbb{1}$ (unity).

(C) Any non-empty set $\mathfrak{F}$ of subsets of a set X is a boolean algebra under the operations meet = intersection, join = union, complement = complement in X, providing $A, B \in \mathfrak{F} \rightarrow A \cap B, A \cup B, X - A \in \mathfrak{F}$. (Such a set $\mathfrak{F}$ is called a *field of subsets* of X) For example, $\mathcal{P}(X)$ is a boolean algebra under the above boolean operations.

(D) Let X be a topological space. Let $\mathcal{C}$ denote the set of all clopen (i.e. closed and open) subsets of X. $\mathcal{C}$ is a field of sets, and hence is a boolean algebra.

(E) Let X be a topological space. Let $\mathcal{R}$ be the set of all closed sets A such that A = closure interior A. Define $A \vee B = A \cup B$, $A \wedge B$ = closure interior $A \cap B$, $-A$ = closure $(X-A)$. Then $\mathcal{R}$ is a boolean algebra. ($\mathcal{R}$ is not usually a field of sets, since, in general, $\wedge$ is not the same as $\cap$).

It can be shown that every boolean algebra is isomorphic to a field of sets (Stone's Theorem : see [4] for details.)

We may define a binary relation on the boolean algebra B by

$$b \leq c \quad \text{iff} \quad b = b \wedge c.$$

(F) For any b, c, $\quad b \leq c \quad \text{iff} \quad b \vee c = c.$

(G) $\leq$ is a partial ordering of B ; $\mathbb{0}$ is the unique minimum of $\leq$ and $\mathbb{1}$ is the unique maximum.

(H) For any b, c, $b \wedge c \leq b \leq b \vee c$.

It is possible to define a boolean algebra as a poset satisfying certain conditions. In this case, $b \vee c$ turns out to be the unique least upper bound of b and c and $b \wedge c$ is the unique greatest lower bound.

2. (Ideals and Filters). Let B be a boolean algebra. A non-empty subset I of B is an *ideal* iff

(a) $b, c \in I \rightarrow b \vee c \in I$;

(b) $b \in I \ \& \ c \in B \rightarrow b \wedge c \in I$.

(A) $I \subseteq B$ is an ideal iff (a) and (b)' hold, where

(b)' $b \in I \ \& \ c \leq b \rightarrow c \in I$.

(B) $\mathbb{0} \in I$ for every ideal I ; if $\mathbb{1} \in I$, then $I = B$.

(C) If $b \in B$, then $\{c \in B \mid c \leq b\}$ is an ideal : it is the *principal ideal generated by* b. (Any ideal not of this form is said to be *non-principal*.)

(D) Let X be an infinite set. Let I be the set of all finite subsets of X. I is a non-principal ideal in the field of sets $\mathcal{P}(X)$.

(E) A *measure* on a boolean algebra B is a function $\mu : B \rightarrow [0,1]$ such that:

(i) $\mu(\mathbb{0}) = 0$, $\mu(\mathbb{1}) = 1$;

(ii) if $b \wedge c = \mathbb{0}$, then $\mu(b \vee c) = \mu(b) + \mu(c)$.

If μ is a measure on B, then $\{b \in B \mid \mu(b) = 0\}$ is an ideal in B.

(F) Let B be a boolean algebra. If I_t, $t \in T$, are ideals in B, so too is $\bigcap_{t \in T} I_t$. Hence if $X \subseteq B$ there is a unique smallest ideal containing X : the ideal *generated* by X.

A non-empty set $F \subseteq B$ is a *filter* iff

(a) $b, c \in F \rightarrow b \wedge c \in F$;

(b) $b \in F \ \& \ c \in B \rightarrow b \vee c \in F$.

(G) In the above definition, (b) can be replaced by

$$(b)' \quad b \in F \ \& \ b \leq c \ \rightarrow \ c \in F.$$

(H) $F \subseteq B$ is a filter iff $\{-b \mid b \in F\}$ is an ideal. (The filter $\{-b \mid b \in I\}$ is the *dual* to the ideal I; and the ideal $\{-b \mid b \in F\}$ is the *dual* to the filter F.)

An ideal in the field of sets $\wp(X)$ is said to be an *ideal* *on* X; similarly *filter* *on* X.

3. (The Order Topology) Let $(X, <)$ be a toset. The *order topology* on X is the topology determined by taking as open subbase all sets of the form $\{x \in X \mid x < a\}$ or $\{x \in X \mid x > a\}$ for some $a \in X$.

(A) The order topology for X is the smallest topology with the property that whenever $a, b \in X$ and $a < b$, then there are neighbourhoods U of a and V of b such that $U < V$ (i.e. $x \in U \ \& \ y \in V \rightarrow x < y$).

(B) If X is connected (with the order topology), then X is complete (as a toset : i.e. every non-empty subset with an upper bound has a least upper bound).

(C) If there are points a, b in X such that $a < b$ and for no c in X is $a < c < b$, we say X has a *gap*. X is connected (with the order topology) iff X is complete (as a toset) and has no gaps.

(D) X is complete (as a toset) iff every closed (in the order topology), bounded subset of X is compact.

Chapter II. The Zermelo-Fraenkel Axioms

In this chapter we develop an axiomatic framework for set theory. For the most part, our axioms will be simple existence assertions about sets, and it may be argued that they are all self-evident "truths" about sets. But why axiomatise set theory in the first place? Well, for one thing, it is well known that set theory provides a unified framework for the whole of pure mathematics, and surely if anything deserves to be put on a sound basis it is such a foundational subject. "But surely", you say, "the concept of a set is so simple that nothing further need be said. We simply regard any collection of objects as a single entity in its own right, and that provides us with our set theory". Alas, nothing could be further from the truth. Certainly, the idea of being able to regard any collection of objects as a single entity forms the very core of set theory. But a great deal more needs to be said about this. Firstly, what is to determine a "collection". In the case of a (small?) finite collection one may simply list the elements of the collection in order to determine it. But what about infinite collections? (or even large finite collections?) Well, we could allow just those collections which are describable by means of a sentence in the English language. But there are only countably many sentences of the English language, so this would not provide us with many sets. Moreover, we would be faced with many collections which are not strictly mathematical, since the expressive power of the English language greatly transcends the realm of mathematics. And we are, after all, looking for a rigorous framework for our set theory. But it would seem that the idea of taking for our "collections" just those collections which are somehow *describable* is quite reasonable. It is just a question of fixing a suitable "language". The "language" must be sufficiently restrictive to allow only the construction of "mathematical" collections, and sufficiently powerful to allow the construction of any set we may require in mathematics. So we commence our study of the concept of a "set" by describing such a language. Later on we shall see whether or not this language helps us in our task of rigorising set theory.

1. The Language of Set Theory

We shall describe a language suitable for, and adequate for, describing mathematical collections. The language will have a precisely fixed set of *symbols* (the "words" of the language) and a rigid *syntax* (= "grammar"). (This will mean that the concept of a "collection describable in the language" will be *rigorously defined*.) As such, it is an example of a *formal language*. Being the language of set theory, let us give it the name LAST. (Admittedly this sounds like the name of a computer programming language. This is no bad thing. Programming languages are also *formal languages*, having the same rigid construction as our own LAST.)

Our language must have a facility for referring to specific sets : we must have a collection of names which we can use to denote sets. Now, at no time shall we be able to refer simultaneously to infinitely many different, specific sets, by name. (This is, after all, a language we are defining, and as such its sentences will just be finite sequences of words of the language.) On the other hand, we could conceivably wish to refer to an arbitrarily large finite number of sets at the same time. (That is, there should be no *a priori* upper bound on the lengths of our sentences, or the number of names of sets which occur in them.) So, what we require is a countably infinite collection of names. Thus, our first requirement is:

(1) *Names (for sets) :* $w_o, w_1, w_2, \ldots, w_n, \ldots$.

These names will be used to denote specific sets. Of course, on one occasion the name w_o may be used to denote one set, on another occasion quite a different set. But this does not matter. During the course of any one description of a set there are enough names to denote all of the sets involved in that description, and it is only a duplication of names occurring in the course of the same description which must be avoided. (Just as the existence of two persons named John Smith only becomes problematical when they live in the same district or work at the same factory, etc.)

Besides referring to specific sets by giving them a (temporary) name, we also wish to refer to *arbitrary* sets. In other words, we need some variables for sets:

(2) *Variables (for sets) :* $v_o, v_1, v_2, \ldots, v_n, \ldots$,

the same argument as for the names leading to our taking a countably infinite collection of such variables.

Next we need to be able to make simple assertions about sets. We need to be able to say that two sets are equal, or that one is an element of the other. So we need :

(3) *Membership symbol :* $\in$;

(4) *Equality symbol :* $=$.

We further need to be able to combine any finite number of assertions, or clauses, to produce one big assertion. So we need

(5) *Logical connectives :* $\wedge$ (and) , $\vee$ (or)

And we clearly require as well a :

(6) *Negation symbol :* $\neg$ (not).

(The use of these symbols is self-evident, but we shall, in any case make this precise when we describe the syntax of LAST.)

Also required are :

(7) *Quantifier symbols* : $\forall$ (for all) , $\exists$ (there exists).

Finally, to serve as punctuation symbols, keeping various clauses apart, we need :

(8) *Brackets* : (,) .

That then is the language LAST. The reader may be surprised to discover (as he will presently) that this simple language is adequate for expressing the most complex of mathematical descriptions. But indeed it is.

As for the syntax, that too is simple. We may build *formulas* (= "clauses", or "phrases", or "sentences") as follows.

(a) Any expression of the forms

$$(v_n = v_m) , \quad (v_n = w_m) , \quad (w_m = v_n) , \quad (w_n = w_m)$$

$$(v_n \in v_m) , \quad (v_n \in w_m) , \quad (w_m \in v_n) , \quad (w_n \in w_m)$$

is a formula of LAST.

(b) If ϕ, ψ are formulas of LAST, so too are

$$(\phi \wedge \psi) \quad , \quad (\phi \vee \psi) .$$

(c) If ϕ is a formula of LAST, so too is

$$(\neg\phi) .$$

(d) If ϕ is a formula of LAST, then

$$(\forall v_n \phi) \quad , \quad (\exists v_n \phi)$$

are formulas of LAST.

No other methods are allowed in the construction of formulas of LAST. Notice that the variables are used in two distinct ways in LAST. If ϕ is a formula of LAST which does not contain a quantifier of the form $\forall v_n$ or $\exists v_n$, then any occurrence of v_n in ϕ is said to be *free* (i.e. v_n denotes an arbitrary set in ϕ). If we now construct the formula $(\forall v_n \phi)$ or $(\exists v_n \phi)$, then all occurrences of v_n in this new formula are said to be *bound*. (Because v_n no longer plays the role of denoting an arbitrary set, of course : it is now an integral part of the quantifier.) A formula which contains no free variables is called a *sentence*. If ϕ is a sentence of LAST, then, once we know which sets any names in ϕ refer to, ϕ can be read as an assertion about sets, and as such will either be true or false.[†] Thus, a sentence actually makes some assertion. A formula which contains any free variables makes no assertion, because there is no meaning available for the free variables. Of course, if we assign specific sets to the free variables we can say whether or not the formula is true *for those assignments* : but on its own the formula has no meaning, since the free variables carry no meaning on their own. We often write $\phi(v_o, \ldots, v_n)$ (etc.) to mean that ϕ is a formula whose free variables (if any) are all amongst the list $v_o, \ldots, v_n$. Given specific sets $a_o, \ldots, a_n$, if we subsequently write $\phi(a_o, \ldots, a_n)$, we mean that ϕ is true when a_i *interprets* v_i in ϕ, for $i = 0, \ldots, n$.

We are now in a position to define the notion of a LAST-*describable collection*. Let $\phi(v_n)$ be a formula of LAST. Once we know which sets the names in ϕ refer to, given any set x we can determine whether or not $\phi(x)$. Hence the collection of all sets x for which $\phi(x)$ is a well-defined collection. And it is clearly a mathematical collection. The only point is : can we obtain all describable mathematical collections in this manner?

[†]There is one possible cause of confusion. Suppose we have a formula $(\forall v_o \phi)$ and we extend this to a formula such as $((v_o = w_o) \wedge (\forall v_o \phi))$. How do we resolve the apparent conflict in the use of v_o? The answer lies in the meaning. In the clause $(\forall v_o \phi)$, v_o is totally "bound" by the quantifier $\forall$, and as such we no longer have any access to it. When we add the conjunct $(v_o = w_o)$, the v_o here is, in a sense, a totally different v_o. The formula $((v_o = w_o) \wedge (\forall v_o \phi))$ thus has exactly one free occurrence of v_o, that occurrence being in the first conjunct. This is, of course, very clear when the meaning of the formula is considered. And we can now go on to form the formula $(\exists v_o ((v_o = w_o) \wedge (\forall v_o \phi)))$. Again, this is unambiguous. One could avoid this complication by altering the syntax somewhat, but there seems no point in doing so when the meaning is so clear.

To put the question a little more precisely: if a collection has any mathematical description, does it have a description in LAST? Of necessity, a formal proof is not possible. The notion of a "mathematical description", though probably well understood, is not a precisely defined notion, whereas the notion of a LAST description is very precise. But by investigating the expressive power of LAST, it soon becomes abundantly clear that it is adequate for any "mathematical description" one could imagine. Part of this procedure has already been carried out for us. We are assuming it is known how all of the concepts of analysis, algebra, etc. can be expressed in terms of sets (e.g. construction of the reals within set theory). So what we must "prove" is that LAST is adequate for expressing any concept of set theory. Now, since LAST is so rudimentary, it is clear that except in the case of very simple assertions, the expression in LAST of any set theoretical assertion will be unbelievably cumbersome, and totally unreadable. And although this is of no consequence to the *fact* of adequacy or otherwise, it would appear to make our task of "proving" adequacy very difficult. But no. We are, after all, only interested in showing that LAST is *capable* of expressing any set theoretical assertion; we do not wish to actually carry out such an expression. So, we are justified in enriching our language by the introduction of abbreviations. For a start, we may introduce the implication symbol $\rightarrow$ as an abbreviation, with

$$(\phi \rightarrow \psi)$$

abbreviating

$$((\neg\phi) \vee \psi)$$

(ϕ, ψ formulas of LAST).

We may then introduce the iff symbol $\leftrightarrow$ as an abbreviation, with

$$(\phi \leftrightarrow \psi)$$

abbreviating

$$((\phi \rightarrow \psi) \wedge (\psi \rightarrow \phi)).$$

Once an abbreviation has been introduced, it may be itself used in order to define new abbreviations. So what we are really doing is this. In set theory, we commence with the very simple notion of sets, equality of sets, and membership of sets, and proceed to develop the whole framework of ordered pairs, functions, partial orderings, etc. from this simple beginning. Our language LAST is adequate for describing the basic part of the development, and hence the whole development. And in order to make this clearer, we expand LAST by introducing abbreviations which correspond to each new development in the set theory. For example, in parallel with the development of set theory carried out in Chapter I, we introduce the following abbreviations into LAST: (we use x, y, z to denote arbitrary names or variables of LAST)

$x \subseteq y$: $(\forall v_n((v_n \in x) \rightarrow (v_n \in y)))$ $(v_n \neq x, y)$;

$x = \cup y$: $(\forall v_n((v_n \in x) \leftrightarrow \exists v_m((v_n \in v_m) \wedge (v_m \in y))))$

$(n \neq m;\ v_n, v_m \neq x, y)$;

$x = \{y\}$: $(\forall v_n((v_n \in x) \leftrightarrow (v_n = y)))$ $(v_n \neq x, y)$;

$x = \{y, z\}$: $(\forall v_n((v_n \in x) \leftrightarrow ((v_n = y) \vee (v_n = z))))$

$(v_n \neq x, y, z)$;

$x = y \cup z$: $x = \cup\{y, z\}$;

$x = (y, z)$: $(\forall v_n((v_n \in x) \leftrightarrow ((v_n = \{y\}) \vee (v_n = \{y, z\}))))$

$(v_n \neq x, y, z)$.

Exercise 1 : *Develop* LAST *further to allow the expression of the following concepts (the first is done for you)*:

(i) x *is an ordered pair;* ($(\exists v_n(\exists v_m(x = (v_n, v_m))))$.)

(ii) x *is a function ;*

(iii) x *is an* n-*ary function on* y;

(iv) x *is a poset ;*

(v) x *is a toset ;*

(vi) x *is a woset ;*

(vii) x *is an ordinal ;*

(viii) x *and* y *are isomorphic wosets ;*

(ix) x *is a group ;*

(x) x *is an abelian group .*

If the reader has faithfully done all of Exercise 1, he will no doubt appreciate how it is that our rudimentary language LAST is capable of expressing very powerful and complex concepts ; basically, because set theory is itself so powerful.

2. <u>The Cumulative Hierarchy of Sets</u>

Having developed our language of set theory to the point we have, it is very tempting to say that a *set* is simply a collection which is describable by a formula of LAST. Thus x will be a set iff there is a formula ϕ of LAST, having one free variable, v_n, say, and sets $a_1, \ldots, a_m$ which the names in ϕ denote, such that x is the collection of all those objects a for which ϕ is true when a interprets the free variable v_n. This definition will certainly provide us with all the sets which are describable in mathematics. (The freedom to refer to any other sets in describing collections overcomes the "handicap" of only using a countable language, and is why there is no bound to the number of sets which we obtain.) And we are clearly unable to describe any non-mathematical collections by formulas of LAST.

So what is wrong with this simple definition of *set*? The answer is immediate: it leads to an inconsistent theory! Indeed, the inconsistency is easily arrived at. Let ϕ be the LAST formula $(\neg(v_0 \in v_0))$. According to the above definition of sets, ϕ defines a set. Thus there is a set x such that

$$x = \{ a \mid a \notin a \} .$$

Now, since x is a set, it must either be the case that $x \in x$, or else that $x \notin x$. If $x \in x$, then x must satisfy the condition imposed by ϕ, that is $x \notin x$. On the other hand, if $x \notin x$, then x must fail to satisfy the ϕ-condition, which means that $x \in x$. So we have a contradiction. Which at once raises the question : "Why, exactly, did the definition fail?" The answer to this question is inherent in the very ideas about a theory of sets which we expressed at the beginning of Chapter I. Fundamental to set theory is the one concept of being able to regard any collection of objects as a single entity. But before we can form a collection of objects, those objects must first be "available" to us. For instance, in our development of naive set theory, we commenced with some initial collection of objects, then considered sets of these objects, then later sets of these sets of objects, and so on. Before we can build sets of sets of objects, we must have the sets of objects out of which to build these sets. The crucial word here, of course, is "build". Naturally we are not thinking of *actually building* sets in any sense, but our set theory should reflect this idea. In the case of our previous definition this was not the case. When we try to form the "set"

$$x = \{ a \mid a \notin a \}$$

the "set" x itself will not be available for consideration as an element. So how can we ever form this set? Indeed, when we build any set u, the set u cannot yet be "available" to us, so it can surely *never* be the case that $u \in u$! Putting these vague considerations into a more precise setting, we see that set theory is essentially hierarchical in nature. We commence with some initial collection, M_0, of objects. We then have a collection, M_1, of sets of members of M_0. Then comes a collection, M_2, of sets of members of $M_o \cup M_1$, and so on. In order to obtain a precise theory now, we must answer three questions:

I. What collection do we take as our initial collection, M_o ?

II. Which "sets" of objects from lower levels of the hierarchy do we take as elements of each new level of the hierarchy ?

III. "How far" does the hierarchy extend ?

Well, since we require our set theory to serve as a foundation for mathematics, it should be as simple and intuitive as possible, with no unnecessary and restrictive assumptions. So, in answer to question I, we commence with *nothing*, i.e. the empty set. Accordingly, we set

$$V_o = \phi ,$$

where V_o denotes the first level of the *set theoretic hierarchy*. Avoiding question II for the moment, let us answer question III. Since our set theory is to have as few restrictions as possible, there should be no point at which we cannot "construct" new sets : thus for each ordinal number α there should be a corresponding level, V_α, in the hierarchy, the members of V_α being sets whose elements all lie in $\bigcup_{\beta<\alpha} V_\beta$. Finally, let us turn to question II. Suppose we have defined the level V_α. Which "sets" of members of V_α are we to take as the members of $V_{\alpha+1}$? Or, to put it another way, since the intention is that $V_{\alpha+1}$ will consist of "all" sets of elements of V_α (this being the "purpose" of the hierarchy), what rules are we to adopt in deciding what is to constitute a "set"? One natural answer is to allow just those collections which are describable in LAST as our "sets". And once a few initial difficulties are overcome, this leads to an extremely rich and powerful theory of sets. But there is another possibility, of a much more general nature. For, when we say that a *collection* can only be said to *exist* if there is some formula of LAST which defines it, we are giving a precise *definition* of the set concept : indeed we are adopting a fundamental axiom of set theory, the *Axiom of*

Constructibility, discussed in Chapter V. But what if we are not so specific, and decide to interpret the word "collection" in the widest possible sense. Given V_α then, we shall say that $V_{\alpha+1}$ will consist of *all* subsets of V_α, without attempting to say what the word "all" really means. This is, of course, much more vague than in the former case, but is nonetheless quite a comprehensible approach. (We all have, do we not, some conception of what the collection of *all* subsets of a set would look like?) Since the Axiom of Constructibility approach will be a sort of "special case", where we actually make the notion of "all subsets" more precise, it is reasonable to take this second notion of set as basic, and see how far we get with it.

We are now able to return to the construction of the set-theoretic hierarchy. In line with our above decision, we take as a basic, undefined (but, hopefully, understood) notion the so-called *unrestricted power set operation* : given any set x, there is a set, $\mathcal{P}(x)$ (the *power set* of x) which consists of all subsets of x. Then, given the level V_α of the hierarchy, we set

$$V_{\alpha+1} = \mathcal{P}(V_\alpha) .$$

This tells us how to go from V_α to $V_{\alpha+1}$. But what do we take for V_α when α is a limit ordinal? (Recall that there are two distinct kinds of ordinals, *successor ordinals* and *limit ordinals*.) One answer might be that we do much the same as above, taking $V_\alpha = \mathcal{P}(\bigcup_{\beta<\alpha} V_\beta)$. Indeed, when we come to investigate the set theoretic hierarchy more thoroughly we shall see that $V_{\alpha+1} = \mathcal{P}(\bigcup_{\beta\leq\alpha} V_\beta)$, so this answer is extremely tempting. But it turns out to be better to take instead the definition

$$V_\alpha = \bigcup_{\beta<\alpha} V_\beta .$$

The reason is, that this reflects more accurately just what is going on at a limit ordinal. When we "form" a limit ordinal, we are really just collecting together all the previous ordinals, without introducing anything new. And this is just what we do in defining V_α above. Of course, this point does not affect the set theory as a whole; it just makes the hierarchy itself more amenable to the demands we shall be making of it. (It is not entirely fatuous to say that, in set theory, it is nice to have time to pause for breath and "collect" oneself every now and then!)

3. Zermelo-Fraenkel Set Theory

We summarise, and at the same time formalise (a little) the discussion of the previous section. In this way we obtain our *theory of sets*.

We take as basic the unrestricted power set operation, $\mathcal{P}(x)$, where $\mathcal{P}(x)$ is the set of all subsets of x. The *cumulative hierarchy of sets* (or the *Zermelo hierarchy*, so named after its inventor) is defined thus:

$$V_0 = \phi \;;$$

$$V_{\alpha+1} = \mathcal{P}(V_\alpha) \;;$$

$$V_\alpha = \bigcup_{\beta<\alpha} V_\beta, \text{ if } \alpha \text{ is a limit ordinal.}$$

A *set* will be an element of some V_α. Because we commence with the empty set, this will mean that, although we place no restriction on the power set operation, only genuine mathematical objects will be allowed as sets. Letting V denote the "collection" of all sets (called the universe of sets), we can express the above "definition of set by the equation:

$$V = \bigcup_\alpha V_\alpha \;.$$

Notice however that this is just a convenient shorthand notation. V is not a *set*, even though it is a well-defined collection. This is because of the "unending" nature of the ordinal numbers. (cf. the situation where W_n is the set of all positive integers less than n, and $W = \bigcup_{n<w} W_n$. W is a well-defined collection of positive integers - indeed it is the set of all positive integers - but it is not a

finite set of integers as is each W_n.)

We have now almost arrived at what is known as *Zermelo-Fraenkel set theory*, named after E. Zermelo and A. Fraenkel, who first formulated and made rigorous this theory. (The intuitive development presented here is essentially due to Zermelo. Fraenkel provided some of the analysis leading to the axiomatisation of the theory, to be described shortly.) There are two principles missing. Firstly, there is the Axiom of Choice, of which more later. Secondly, and more fundamentally, since we have not described the power set operation at all, how can we be sure that all the sets we require in mathematics will appear in our collection, V, of all sets? As a fundamental axiom we need the following principle:

Axiom of Subset Selection

Let x be a set, and let $\phi(v_n)$ be a formula of LAST (which may, as usual, refer to sets by name). Then amongst the sets in $\mathcal{P}(x)$ appears the set of all those members a of x for which $\phi(a)$.

Of course, it is an immediate consequence of *an understanding* of what $\mathcal{P}(x)$ means that the above principle is *true*. But since we took the power set operation as totally undescribed *in our theory*, the above principle must be included, to reflect within the theory our intuition about $\mathcal{P}(x)$. (A similar remark will apply to the Axiom of Choice, when we come to discuss it.)

In essence, *Zermelo-Fraenkel set theory* can be summarised as the theory of sets with the assumptions:

I. $V = \bigcup_{\alpha} V_\alpha$;

II. Axiom of Subset Selection ;

III. Axiom of Choice (see later).

That is, sets occur in a hierarchy which commences with the empty set and proceeds via the formation of "new" sets from "old" objects; and in the formation of sets, all mathematical constructions of sets are allowed (this being formalised by the language LAST). Plus the Axiom of Choice.

Exercise 1 : *Show that for any ordinal* α, $V_\alpha = \bigcup_{\beta<\alpha} \mathcal{P}(V_\beta)$.

Exercise 2 : *Show that if* $\alpha < \beta$, *then* $V_\alpha \subseteq V_\beta$. *(This explains the use of the phrase* "cumulative hierarchy of sets" *to describe the* V_α*-hierarchy.)*

Exercise 3 : *Check that:*

(i) $V_o = \phi$;

(ii) $V_1 = \{\phi\}$;

(iii) $V_2 = \{\phi, \{\phi\}\}$;

(iv) $V_3 = \{\phi, \{\phi\}, \{\{\phi\}\}, \{\phi, \{\phi\}\}\}$.

Exercise 4 : *What are :*

(i) V_4 , (ii) V_5 ?

Exercise 5 : *In general, how many elements will* V_n *have, where* n *is a positive integer?*

Exercise 6 : *Show that if* $y \in V_\alpha$ *and* $x \in y$, *then* $x \in V_\alpha$. *(A set* M *is said to be transitive if* $x \in M \rightarrow x \subseteq M$. *Thus we can rephrase this exercise by saying that each* V_α *is transitive. Why is the word "transitive" used here?)*

4. Axioms for Set Theory

We have now developed a simple and (we hope) adequate theory of sets. (With reference to the parenthesised comment in the previous sentence, let us remark that the system developed is certainly adequate for the development of all of contemporary mathematics. But this question of adequacy, together with that of consistency, will be considered later.) But the description of the theory depends upon the construction of the cumulative hierarchy of sets, the V_α-hierarchy, and this, in turn, depends upon the ordinal number system. We are thus assuming a considerable amount of "set theory" in order to define our *set theory*. There is, of course, no real dilemma here. What we have done is to analyse what we mean by the concept of a "set", and our set theory has been the result of this analysis. We have not really *defined* set theory. This will be the next step, in line with centuries of mathematical development. By analysing still further, we shall isolate those fundamental assumptions about sets which are implicitly required in order to obtain the set theory developed above, and then, by taking these assumptions as the *axioms* of set theory, we shall turn the whole process round, obtaining a well-defined set theory, based on a set of axioms. Now, since we have developed our theory of sets thus far guided by our intuition, it is to be expected that the axioms we obtain will be very intuitive. And indeed this will be the case. But since the concept of a set is so intuitive (is it?), one might also expect the axioms to form a neat collection of about three statements. In fact we shall obtain what can only be called a motley collection of some nine statements. To the beginner who is "worried" by this we can really only make two comments. Firstly, each of the axioms is certainly a principle about sets which cannot be disputed. Secondly, the system of set theory which the axioms give us is really just the intuitive system described in §3 : and that is intuitively acceptable.

Let us commence then, an enumeration of the assumptions which we implicitly used in constructing the V_α-hierarchy of sets. Well, for a start we took the power set operation as basic. So we are assuming that for any set x, there is a set which consists of all subsets of x (i.e. the power set of x). Formulating this as an axiom of set theory, we have :

Power Set Axiom

If x is a set, there is a set which consists precisely of all the subsets of x.

Exercise 1 : *Write down a sentence of* LAST *which expresses the power set axiom.*

The power set axiom allows us to pass from V_α to $V_{\alpha+1}$ in the construction of the cumulative hierarchy of sets. What about the definition of V_α when α is a limit ordinal? Well, in this case we have

$$V_\alpha = \bigcup_{\beta<\alpha} V_\beta ,$$

so we must be able to form the union of any collection of sets:

Axiom of Union

If x is a set, there is a set whose members are precisely the members of the members of x (namely, $\cup x$).

Exercise 2 : *Express the axiom of union in the language* LAST.

The axiom of union allows us to obtain V_α for limit α as

$$V_\alpha = \cup\{V_\beta \mid \beta < \alpha\} .$$

But wait a moment. How do we know that $\{V_\beta \mid \beta < \alpha\}$ is a set? Well, $\alpha = \{\beta \mid \beta < \alpha\}$ is a set. (For the time being we are taking the ordinal numbers as basic. Later we shall see what assumptions are needed for their construction.) And one can obtain $\{V_\beta \mid \beta < \alpha\}$ from the set $\{\beta \mid \beta < \alpha\}$ by *replacing* each element β of α by the set V_β. This leads us to the *Axiom of Replacement*. It is perhaps one of the least appreciated axioms of set theory. And yet it is undoubtedly one of the most powerful axioms. The main reason why the non-expert finds it hard to appreciate the axiom of replacement is that it is rarely required in most areas of mathematics. It is predominantly an axiom for the set theorist. There are, however, several instances where it is known for certain that it is necessary in order to obtain results about the real line, for example, so it should not be ignored. Roughly speaking, what the axiom of replacement says is that if we have a set x, and we *replace* each element a of x by a new set a', then the collection of all a' so obtained is a set. The immediate question is : what is to determine a "replacement"? If x is finite, we can list the elements a of x and alongside them the new sets a', and in this manner we can say exactly what the replacement procedure is. But what in the general case? Well, the answer should, by now, be obvious. We allow any replacement procedure which can be described by a formula of LAST. Thus, what the axiom of replacement will say is : if x is a set and F is a function from x to sets which is definable by a formula of LAST, then $\{F(a) \mid a \in x\}$ is a set. But is this not a circular definition? A function, if it is at all to exist, is a *set* of ordered pairs, so are we not almost assuming the conclusion of the axiom in order to state it? Well no. True, a function, as we have (and shall continue to) defined it is a *set* of ordered pairs. That tells us what kind of set-theoretical *object* a function is. But we were not using functions as objects in formulating the axiom of replacement, rather as "rules of replacement". In other words, any circularity arising in our first formulation of the axiom only arose because of the special meaning we have attached to the word "function". We overcome this by referring not to *the function* but to the rule which defines it.

Axiom of Replacement

Let $\phi(v_n, v_m)$ be any formula of LAST (which may refer by name to any finite number of specific sets), and let x be a set. Suppose that for each set a there is a unique set b such that $\phi(a, b)$. Then there is a set y consisting of just those b such that $\phi(a, b)$ for some a in x.

Exercise 3 *It is not possible to transcribe the axiom of replacement as stated above to a sentence of* LAST, *since* LAST *has no facility for handling formulas of*

LAST *themselves. This difficulty may be overcome by regarding the axiom of replacement as an axiom schema, each formula* ϕ *giving rise to a specific instance of this schema. Given any formula* ϕ *of* LAST *with free variables* v_n *and* v_m *only, write down the sentence of* LAST *which expresses the* ϕ*-instance of the axiom of replacement. (The Axiom of Replacement thus says that all the sentences of* LAST *of the kind you have (we hope) written down are true.)*

We now have the axioms we need in order to construct the cumulative hierarchy of sets, given the ordinal number system. We turn now to the question of what we need in order to construct the ordinals. It turns out that only a few very simple requirements are left to be formulated as axioms. These are as follows.

Null Set Axiom

There is a set which has no members. (This set being denoted by $\emptyset$.)

Axiom of Infinity

There is a set x such that $\emptyset \in X$ and such that $\{a\} \in x$ whenever $a \in x$.

Some comment is called for concerning this last axiom. Axioms such as the Power Set Axiom, although providing us with *new* sets, require the existence of sets before they can function, and as such do not in themselves guarantee that our set theoretic universe, V, will be non-trivial. Only two of our axioms do this: the Null Set Axiom and the Axiom of Infinity. Taken with the Null Set Axiom, the other axioms of set theory (leaving aside the Axiom of Infinity for the moment) allow us to construct many finite sets. But without the Axiom of Infinity we are unable to pass into the realm of the transfinite, and this is, after all, what set theory is all about. Now, in order to obtain all the infinite sets we need, it suffices that we commence with just one infinite set. The precise nature of this set turns out to be quite irrelevant, so we have some freedom in the way we formulate the Axiom of Infinity . (But notice that the notion of "infinite" is not itself a basic notion in our theory.) The formulation chosen has the advantage of being easy to state.

The reader should bear in mind that although in our subsequent development we shall be able to construct sets which are in every way imaginable immeasurably larger than the set of natural numbers (say), no further "axioms of infinity" will be required to do this; the one leap provided by the Axiom of Infinity being sufficient. As such, the Axiom of Infinity is an extremely powerful assumption.[†]

Of course, since the Axiom of Infinity guarantees the existence of at least one set, we can *prove* the Null Set Axiom by a simple application of the Axiom of Subset Selection (e.g. given some set a, we have $\emptyset = \{x \in a \mid x \neq x\}$), so we could omit the Null Set Axiom from an enumeration of the axioms of set theory if we wished. However, in view of its fundamental nature it is usual to include it as an axiom in its own right.

And one final remark. Our formulation of the Axiom of Infinity requires the existence of the operation $a \to \{a\}$. Many texts include a "Pairing Axiom" in their axiomatisation of set theory, guaranteeing the existence of the unordered pair $\{a, b\}$ of any sets a, b. A special case of this then provides singletons, of course. However, all finite sets can be easily obtained by applying the Axiom of Replacement to the sets $\mathcal{P}(\emptyset)$, $\mathcal{P}\mathcal{P}(\emptyset)$, $\mathcal{P}\mathcal{P}\mathcal{P}(\emptyset)$, etc. (*Exercise: Check this.*). Consequently we shall not regard the "Pairing Axiom" as a fundamental axiom.

<u>Exercise 4</u>: *Express the above two axioms in LAST.*

Have we forgotten anything? Well, we have not mentioned the Axiom of Subset Selection in the above list, but the adoption of this principle has long since been acknowledged. Anything else? The answer is "Yes", but the remaining axiom is so

---

[†]Indeed, the knowledge that Zermelo-Fraenkel set theory is not able to resolve all the questions about sets which may be formulated in the theory led various people to consider extensions of the theory obtained by introducing additional "axioms of infinity", trying to mimic at a higher level the jump from the finite to the infinite provided by the Axiom of Infinity. In no case could it be said that the attempt came anywhere near to achieving its aim. (See III.10 for further details.)

very fundamental that it could easily be forgotten. We are, after all, considering a theory of *sets*, and a set is just a collection of objects, so we have an axiom which reflects this fact within the theory, namely:

Axiom of Extensionality.

If two sets have identical elements, then they are equal.

The converse to the above assertion is also valid, of course, but that need not be included here since it is a theorem of logic.

Exercise 5: *Express the Axiom of Extensionality in* LAST.

Our analysis is now complete. The following collection of axioms suffices for the construction of the ordinal number system and the cumulative hierarchy of sets:

1. Axiom of Extensionality
2. Null Set Axiom
3. Axiom of Infinity
4. Power Set Axiom
5. Axiom of Union
6. Axiom of Replacement
7. Axiom of Subset Selection.

The axioms of Zermelo-Fraenkel set theory then consist of the above seven statements, together with the following two:

8. $V = \bigcup_{\alpha} V_{\alpha}$
9. Axiom of Choice.

Leaving aside the formulation of the Axiom of Choice for the time being, the above description of Zermelo-Fraenkel set theory, whilst accurate, is not in its most concise form. The problem is the formulation of axiom 8. In order to *state* this axiom, we have had to assume a fair development of the theory based on the other axioms, at least as far as the construction of the ordinal number system and the cumulative hierarchy of sets. How much nicer it would be if we could replace statement 8 by a more basic assertion. This turns out to be quite easy. In the presence of axioms 1 to 7, statement 8 is equivalent to the fact that the binary relation of set membership ($\in$) is well-founded. So we may replace 8 by the more fundamental axiom:

Axiom of Foundation

$\in$ is a well-founded relation.

A more explicit way of expressing the above axiom is :

For every set x there is a set $a \in x$ such that $a \cap x = \emptyset$.

Exercise 6: *Assuming axioms 1 to 7 in the above list, prove that the Axiom of Foundation as above is equivalent to the equality*

$$V = \bigcup_{\alpha} V_{\alpha} .$$

Exercise 7: *Show that the two versions of the axiom of foundation given above are equivalent. (In other words, what does it mean to say that "$\in$ is well-founded".)*

5. Summary of the Zermelo-Fraenkel Axioms

We list the axiom system evolved above.

1. Axiom of Extensionality.

If two sets have the same elements, then they are equal.

2. Null Set Axiom.

There is a set, ϕ, which has no members.

3. Axiom of Infinity

There is a set x such that $\emptyset \in x$ and such that $\{a\} \in x$ whenever $a \in x$.

4. Power Set Axiom

If x is a set, there is a set, $\mathcal{P}(x)$, consisting of all subsets of x.

5. Axiom of Union

If x is a set, there is a set, $\cup x$, consisting of all elements of all elements of x.

6. Axiom of Replacement

Let x be a set, and let F be a "function" defined on x by a formula of LAST (which may refer by name to existing sets). Then $\{F(a) \mid a \in x\}$ is a set.

7. Axiom of Subset Selection

Let x be a set, and let $\phi(v_n)$ be a formula of LAST (which may refer by name to existing sets). Then there is a set consisting of just those a in x for which $\phi(a)$.

8. Axiom of Foundation

If x is a set there is an $a \in x$ such that $a \cap x = \emptyset$.

9. Axiom of Choice (See later)

The theory whose axioms are 1 to 8 above is usually denoted by ZF. If we add axiom 9, we denote the resulting theory by ZFC. This is at slight variance with the fact that "Zermelo-Fraenkel set theory" has all nine axioms as its basic assumptions, but the nomenclature is now standard.

<u>Exercise 1</u> : *The nine axioms listed above are not all independent. For instance, the null set axiom may be deduced from the other ZF axioms, and so too may the axiom of subset selection. (This last part requires clever use of the axiom of replacement. Given a set* x *and a formula* ϕ *of* LAST, *consider the "function"* F *defined by*

$$F(a) = \begin{cases} \{a\}, & \text{if } \phi \text{ is true for } a \\ \emptyset, & \text{otherwise.} \end{cases}$$

Now consider the set $\cup\{F(a) \mid a \in x\}$.) *Nevertheless, it is convenient to state them all, as they explain more fully what the theory is about.*

<u>Exercise 2</u> : *Examine the development of the ordinal numbers in* I.7 *and see how the various axioms are used, paying particular attention to the use of the Axiom of Replacement in the proof of* I.7.11.

6. Classes

Sooner or later, in any course on set theory, the instructor finds it necessary to introduce the concept of a proper class. And this moment he usually dreads. For with virtual certainty, one of two things will occur. Either his students, hitherto having found the development of set theory entirely plausible and logical, will suddenly gain the uneasy feeling that a *sleight-of-hand* method of introducing some very dubious notion is being employed. Or else the students will accept the new notion but fail totally to understand what exactly is being done. For us, this moment has come. We urge the student to reflect at some length on what is going on.

Basically, the idea is very simple. Sets are, from the point of view of *set theory*, completed entities (one might even say *points* in the space of all sets). And our set theory tells us how to handle these entities. Now, as we know, a set is a collection of objects (those objects also being sets, in fact). Does it follow that any collection of objects is a set? Well, before we can answer this, we should ask ourselves what is meant by the words "collection" and "object" here. By "object" we clearly mean "set". And by "collection", we mean, of course, "collection determined by a formula of LAST (where the names in that formula may refer to any specific sets)". (What else could constitute a *well-defined collection* of sets in our context?) Having now clarified our question, we see that the answer is "no". For instance, the collection, V, of *all* sets is not itself a set. (If it *were*, then by the axiom of subset selection,

$$\{ x \in V \mid x \notin x \}$$

would be a set, and we have already seen what happens then!) And yet V is a well-defined collection. Indeed, if $\phi(v_o)$ is the formula " $v_o = v_o$ " of LAST, then V is just the collection of all sets x for which $\phi(x)$ is true; i.e.

$$V = \{ x \mid x = x \} .$$

Hence there are collections of sets, definable by formulas of LAST, which are not sets. Since these collections are not sets, the Zermelo-Fraenkel axioms do not tell us how to handle them. They are somehow "too big" : they are not "completed collections", as are sets. Nonetheless, it would be a nuisance if we could not discuss such collections. There are two ways to do this. One is to enlarge the axiom system to facilitate the handling of such collections. (We are then, of course, not really doing *set* theory but *class* theory, which includes set theory as a subsystem.) This can be done. The most common system is due to Bernays and Gödel, and is described in Monk [5]. The main disadvantages (as we see it) with this system are that it is not *necessary* to introduce it, and that it results in a loss of the intuitive naturalness of Zermelo-Fraenkel set theory. As we view matters, set theory is based on the idea of iteratively constructing new sets from old ones (so in set theory one is always climbing upwards), whereas in class theory the "universe" of sets is a completed whole (so one looks downwards at the universe of sets from a high vantage point). And when we said above that it was not necessary to formalise a theory of "big collections", we were not just expressing a vague idea. One can *prove* that any result about *sets* which is provable in Bernays-Gödel class theory is already provable in Zermelo-Fraenkel set theory.

The other way of dealing with "big collections" is as follows. We introduce the notion of a "class" as a convenient *abbreviational* device. Given any formula $\phi(v_n)$ of LAST, whose names refer to specific sets, the collection

$$\{ x \mid \phi(x) \}$$

of all x for which $\phi(x)$ is said to be a *class*. Now, all sets are classes. Indeed, if a is a set, the LAST formula "$v_o \in w_o$" defines the class a when w_o denotes a; i.e.

$$a = \{ x \mid x \in a \} .$$

But, as we saw above, not all classes will be sets. For instance, V is a class which is not a set. Such classes will be called *proper classes*. Since proper classes are not sets, we are not able to handle classes as we do sets. For instance, we cannot ask ourselves if one class is a member of another. This has no meaning. A proper class is an "uncompleted collection", and hence is never available for being *in* any other collection. It is not just false to write "$V \in V$", it is meaningless. The statement "$V \notin V$" is likewise meaningless. So what, you may ask, is the point of introducing classes? Well, classes *are* collections, and hence will exhibit many of the properties of sets. And providing we exercise a little care, we can handle classes quite often just as if they were sets. Indeed, the only thing we must never do is treat a proper class as a "completed whole", or "point". But surely, you say, what we have now done is enlarged our theory to incorporate classes? Well no, because we shall only use them as abbreviations. If A is some class, then there will be a LAST formula $\phi(v_o)$ such that

$$A = \{ x \mid \phi(x) \}.$$

In discussing the "class" A, we are simply avoiding the explicit mention of ϕ. If challenged by a "purist", we could always stop referring to A and deal with ϕ instead. For instance, if I wrote

$$a \in A$$

and you were upset by my use of the symbol A to denote something which, by my own admission, does not really *exist* (in the sense of set theory), I could instead write

$$\phi(a).$$

The two statements clearly have the same meaning, but in the second no use is made of classes.

Again, if

$$A = \{ x \mid \phi(x) \}$$

$$B = \{ x \mid \psi(x) \},$$

and if I wrote

$$A = B,$$

then I could always replace this by the totally harmless statement

$$\forall x \ (\phi(x) \leftrightarrow \psi(x)) .$$

Likewise,

$$A \subseteq B$$

can be replaced by

$$\forall x(\phi(x) \to \psi(x)).$$

Exercise 1. *Let* A, B, ϕ, ψ *be as above. Let* C *be the class* $A \cup B$. *Express the assertion*

$$x \in C$$

as a sentence in set theory.

Exercise 2. *As above, but now take* $C = A \cap B$.

Admittedly, our above examples only give simple instances, but already it should be getting clearer. In fact complex assertions about classes will be just combinations of very basic assertions (just as with assertions about sets), so the examples above are really all that we need. All uses of classes reduce to statements of the form

the set a is a member of the class A

(For instance "the class A equals the class B" can be reduced to "for all sets a, a is a member of A iff a is a member of B".)

Now, so far, it may not be apparent that there is a great deal to be gained by introducing classes. Indeed, in the sense of achieving a stronger theory, there is no gain at all: they are just abbreviations. But do they help us to understand things better, and do they ever help clarify various concepts? The answer is an emphatic "yes". For instance, consider the statement of the axiom of replacement given in II.4. This ran as follows:

Let $\phi(v_n, v_m)$ be any formula of LAST and let x be a set. Suppose that for each set a there is a unique set b such that

$\phi(a, b)$. Then there is a set y consisting of just those b such that $\phi(a, b)$ for some a in x.

Quite a mouthful, and difficult to read. The difficulty can be totally eliminated by introducing classes. The formula ϕ above clearly defines a "class function". That is, if

$$F = \{(a,b) \mid \phi(a,b)\} ,$$

then, except for the fact that it is not a set, the class F has all the appearances and properties of a function. And, referring to the class F, what the axiom of replacement says is that for any set x, $\{F(a) \mid a \in x\}$ is a set. Simple, concise, and totally unambiguous. Indeed, we have already expressed the axiom of replacement in this manner in II.5. And I'll bet you lost no sleep over our usage then, even though we had not discussed classes as such! Which fact should emphasise that there is no question of any "dubious practice" going on when we introduce classes. Indeed, mathematicians automatically think along the lines which lead to classes (unless they are category theorists, perhaps).

The introduction (and elimination) of classes, though natural, does admittedly take some getting used to. If the reader is still a little puzzled, let him read on, for in the next two sections he will find examples abundant. We finish by summarising what has been said above.

1. Officially, classes are just abbreviations. Their use can always be eliminated by replacing them by the formulas of LAST which define them.
2. Classes may be thought of as "big collections".
3. Classes can be handled as sets except that the class is never a completed whole to be a possible member of anything else, so, e.g. $\mathcal{P}(A)$ has no meaning for A a proper class.
4. All sets are classes. Some classes (proper classes) are not sets.

Exercise 3. *Let* On *be the class of all ordinals. Prove that* On *is a proper class. (Hint. Show that if* On *were a set, it would be an ordinal, whence we would have* On $\in$ On.)

Exercise 4. *Show that the assertions* (a), (b) *below are equivalent:*

(a) $(\forall x \in V)(\exists y \in On)(\exists f \in V)[f : x \leftrightarrow y]$

(b) *Every set can be well-ordered.*

Exercise 5. *Let* $A = \{ x \mid (\exists y \in On)(\exists f)(f : x \leftrightarrow y)\}$. *Show that* (b) *above can be expressed thus:*

$$V = A.$$

7. Set Theory as an Axiomatic Theory.

We were led to our system Zermelo-Fraenkel set theory by way of the Zermelo hierarchy of sets. Now that we have formulated this system, we can take the axioms ZFC as basic and develop set theory rigorously from these axioms. And providing the axiom system ZFC is consistent, we can be sure that anything we prove in our set theory is meaningful. Indeed, if we "believe" the axioms ZFC, we can conclude that anything proved from them is "true". (The reader who so wishes is permitted to delete the quotation marks from the last sentence.) Now, it is a consequence of a classical theorem of logic, due to Gödel, that we cannot hope to *prove* that ZFC is consistent. In order to prove the consistency of ZFC, one would need to carry out the proof itself in a theory even stronger than ZFC, whose own consistency would be even more in doubt, of course. With a foundational subject like set theory, one is forced to make an *assumption* of consistency somewhere along the line. In fact, we *can* go *one* step beyond assuming the system ZFC is free of contradictions. Another theorem of Gödel shows that if ZFC is an inconsistent theory, then so too is ZF. So one simply needs to assume that ZF is consistent in order to be sure that

ZFC is consistent. We shall assume throughout that ZF is consistent (for otherwise, there would be no point in our writing this book).

So, granted our consistency assumption, anything we prove from the axioms ZFC will be a meaningful assertion about sets. But when we formulated the system ZFC, did we miss anything fundamental? More precisely, we formulated the ZFC axioms in an attempt to make precise the basic assumptions about sets which we must implicitly make when we wish to develop a theory of sets in the manner outlined in section 2. Do the ZFC axioms do this? This reduces at once to the more precise question : assuming only the axioms of ZFC, can we define the Zermelo hierarchy, V_α, $\alpha \in$ On.

Well, how did we define the Zermelo hierarchy? To commence we set $V_0 = \emptyset$. Then, given V_α we set $V_{\alpha+1} = \mathcal{P}(V_\alpha)$. And if λ was a limit ordinal and V_α was defined for all $\alpha < \lambda$, we set $V_\lambda = \bigcup_{\alpha<\lambda} V_\alpha$. It is easily seen that this definition can be expressed more simply as:

$$V_\alpha = \bigcup_{\beta<\alpha} \mathcal{P}(V_\beta) . \qquad (\forall \alpha)$$

(<u>Exercise 1</u>. *Prove the equivalence of the two versions of the definition of the Zermelo hierarchy given above.*)

Thus, in order to define V_α, we first need to have defined all the sets V_β, $\beta < \alpha$. In fact, we need more. In order to define V_α we need to have available the sequence (or function) $\langle V_\beta | \beta < \alpha \rangle$ which assigns to each ordinal $\beta < \alpha$ the corresponding set V_β. For, letting f denote this function, we actually define V_α as

$$V_\alpha = \bigcup\{ \mathcal{P}(f(\beta)) \mid \beta < \alpha \} .$$

(By the axioms of power set, replacement, and union, this is an admissible definition of a set.) Such definitions are sometimes referred to as *definitions* by *induction*. More correctly they are definitions by *recursion*. (*Induction* is a method of *proof*, not of construction.)

Letting $f : On \to V$ (use of class notation!) be the "function" $f(\alpha) = V_\alpha$, we define $f(\alpha)$ in terms of $f \restriction \alpha$ (i.e. in terms of $\langle f(\beta) \mid \beta < \alpha \rangle$). Indeed, we have

$$f(\alpha) = \cup\{ \mathcal{P}((f \restriction \alpha)(\beta)) \mid \beta < \alpha \} .$$

That such definitions are possible (on the basis of ZF) is a consequence of the *recursion principle*.

8. The Recursion Principle

Although basically simple in concept and application, the recursion principle is often not fully appreciated. It plays a central role in set theory, and its importance cannot be over-emphasised. Intuitively, all it says is that, in the ZF system, it is possible to define functions by recursion; hence it can be applied without being fully understood. But unfortunately, from a set theoretical point of view there are some awkward complications arising when one tries to state and prove the recursion principle. It is these complications which usually prevent students from gaining a mastery, and consequent understanding, of the principle. (Another difficulty comes from the fact that the idea of defining a function by a recursive procedure is highly intuitive, whereas proving on the basis of the axiom system ZF that a certain function "exists" is not at all intuitive. So one is proving "the obvious" in a non-obvious manner!)

Before we start, let us therefore warn the casual reader that he may well find the following discussion rather hard to follow. In this case, let us reassure him that if he contents himself with the knowledge that recursive definitions are always possible, he can read the rest of the book without any further loss. But the reader interested in set theory *per se* should prepare himself for some hard work. He must grasp this very central concept.

Now, starting with the ZF axioms (we do not need the Axiom of Choice here), the ordinal number system can be developed as in I.7. Assuming this development from now on, we first state a simplified recursion principle.

8.1 Theorem (Recursion on an Ordinal)

Let $h : On \times V \to V$ be a "class function". Let λ be an ordinal. Then there exists a unique function $f : \lambda \to V$ such that for every $\alpha \in \lambda$:

$$f(\alpha) = h(\alpha, f \upharpoonright \alpha). \qquad \square$$

We shall prove 8.1 presently, but let us first see how the use of classes can be eliminated from the above statement.[†] Well, strictly speaking we are developing the theory whose axioms are ZF. (Choice is still not required) As it stands, 8.1 is an assertion which implicitly involves a universal quantifier, $\forall h$, ranging over proper classes. This is not possible in the ZF system. But now fix our attention on a single (but arbitrary) h. Let $\phi(v_o, v_1, v_2)$ be that formula of LAST which defines h. That is, for $\alpha \in On$, $x, y \in V$,

$$h(\alpha, x) = y \quad \text{if} \quad \phi(\alpha, x, y) .$$

Starting with this formula ϕ, we may prove, on the basis of ZF, that there exists a unique function $f : \lambda \to V$ such that for every $\alpha \in \lambda$,

$$\phi(\alpha, f \upharpoonright \alpha, f(\alpha)).$$

In other words, 8.1 as stated really represents a schema of theorems of ZF. For each h, there is a corresponding ZF theorem. The only thing which ZF cannot prove is 8.1 *as stated*. (The reason for stating 8.1 as we did is, however, clear. Reformulation of the theorem along the lines indicated above to eliminate the use of classes results in a rather complex assertion.)

[†]This discussion concerns a rather subtle point, and the reader may well find it difficult to see what is going on. In which case he should perhaps postpone reading it in detail until later. Indeed, for the casual reader it can safely be ignored. Only the intending mathematical logician needs to *master* the point discussed (some time or other).

Before proving 8.1, it is perhaps worth our while seeing how this helps us to define the Zermelo hierarchy. In fact, all 8.1 tells us is that for every λ, the hierarchy $\langle V_\alpha \mid \alpha \in \lambda \rangle$ exists (as a function with domain λ). When we have proved 8.1 we shall see how it can be extended to give the full hierarchy (which is, of course, a class "function", with "domain" On). So, fixing λ, consider $h : \mathrm{On} \times V \to V$ defined by

$$h(\alpha, x) = \begin{cases} \bigcup_{\xi \in \mathrm{dom}(x)} \mathcal{P}(x(\xi)), & \text{if } x \text{ is a function,} \\ \emptyset, & \text{otherwise.} \end{cases}$$

By the axioms of power set, replacement, and union, h is a well-defined function. By 8.1, there is a function $f : \lambda \to V$ so that

$$f(\alpha) = h(\alpha, f \restriction \alpha)$$

for all α. By definition of h, this means that for all $\alpha \in \lambda$,

$$f(\alpha) = \bigcup_{\xi<\alpha} \mathcal{P}(f(\beta)) .$$

Indeed, 8.1 tells us that this f is unique. Clearly, f is what we want: f is our sequence $\langle V_\alpha \mid \alpha < \lambda \rangle$. In other words, $\langle V_\alpha \mid \alpha < \lambda \rangle$ is the unique function which 8.1 guarantees us when we define h as above.

We turn now to the proof of 8.1. Let $\lambda \in \mathrm{On}$, and let $h : \mathrm{On} \times V \to V$. Using only the axioms of ZF, we prove that there is a unique function $f : \lambda \to V$ such that $f(\alpha) = h(\alpha, f \restriction \alpha)$ for all $\alpha \in \lambda$. We first prove uniqueness.

8.2 <u>Lemma</u>

Let $\mu \leq \lambda$. Suppose $f_i : \mu \to V$, $i = 1, 2$, are such that for all $\alpha \in \mu$, $f_i(\alpha) = h(\alpha, f_i \restriction \alpha)$. Then $f_1 = f_2$.

Proof: By induction on μ. (By I.7.1, we may prove results by induction on ordinals.) For $\mu = 0$ it is trivial. Let $\mu \leq \lambda$, and suppose the results holds for all $\mu' < \mu$. Thus, for $\mu' < \mu$, $f_1 \upharpoonright \mu' = f_2 \upharpoonright \mu'$. If $\lim(\mu)$, then it follows at once that $f_1 = f_2$. Otherwise, if $\mu = \nu+1$, then we have, by induction, $f_1 \upharpoonright \nu = f_2 \upharpoonright \nu$. Hence

$$f_1(\nu) = h(\nu, f_1 \upharpoonright \nu) = h(\nu, f_2 \upharpoonright \nu) = f_2(\nu).$$

Thus, $f_1 = (f_1 \upharpoonright \nu) \cup \{(\nu, f_1(\nu))\} = (f_2 \upharpoonright \nu) \cup \{(\nu, f_2(\nu))\} = f_2$. □

Now let M be the class

$$\{f \mid (\exists \mu \leq \lambda)[f : \mu \to V \ \& \ (\forall \alpha \in \mu)(f(\alpha) = h(\alpha, f \upharpoonright \alpha))]\}.$$

8.3 Lemma

Let f, g $\in$ M. Let $\mu = \mathrm{dom}(f)$, $\nu = \mathrm{dom}(g)$, and suppose $\mu < \nu$. Then $f = g \upharpoonright \mu$.

Proof: For all $\alpha \in \mu$, we have :

$$f(\alpha) = h(\alpha, f \upharpoonright \alpha)$$

$$g(\alpha) = h(\alpha, g \upharpoonright \alpha).$$

So, by 8.2, $f = g \upharpoonright \mu$. □

In order to prove 8.1, it suffices to show that for some $f \in M$, $\mathrm{dom}(f) = \lambda$. Let

$$A = \{ \mu \mid (\exists f \in M)(\mathrm{dom}(f) = \mu) \}.$$

We must show that $\lambda \in A$. Suppose not. Then $(\lambda+1) - A \neq \emptyset$. Let μ be the least element of this set. (Thus $\mu \leq \lambda$). Thus, for each $\nu < \mu$ there is an $f \in M$ with $\mathrm{dom}(f) = \nu$. By 8.3, we may define a class function H on μ by taking for each $\nu < \mu$, $H(\nu)$ to be that unique $f \in M$ with $\mathrm{dom}(f) = \nu$. By the axiom of replacement, $H[\mu]$ is a set. Let $f = \cup H[\mu]$. Using 8.3, it is easily seen that f is a function. Moreover, for each $\nu < \mu$, $f \upharpoonright \nu = H(\nu)$, so for all $\nu < \mu$, we have

$$(\forall \alpha \in \nu)(f(\alpha) = h(\alpha, f \upharpoonright \alpha)).$$

If $\lim(\mu)$, then this implies that $f \in M$ and $\mathrm{dom}(f) = \mu$, contrary to the choice of μ. So we must have $\mu = \nu+1$, some ν. Now set

$$f' = f \cup \{ (\nu, h(\nu, f)) \} .$$

Then $f' \in M$ and $\mathrm{dom}(f') = \mu$, a contradiction. This completes the proof of 8.1.

Exercise 2. *Write out a proof of* 8.1 *which avoids all use of classes.*

We are now ready to state the full ordinal recursion principle. This will provide us with the Zermelo hierarchy $\langle V_\alpha \mid \alpha \in On \rangle$ in the manner analogous to the way we obtained its initial parts from 8.1

8.4 Theorem (Ordinal Recursion)

Let $h : On \times V \to V$. Then there exists a unique $f : On \to V$ such that for every $\alpha \in On$,

$$f(\alpha) = h(\alpha, f \upharpoonright \alpha). \quad \square$$

Clearly, 8.4 is a sort of "limiting" version of 8.1. But now when we come to examine what is really meant by the usage of the classes h, f we must be more careful than before.†

8.4 is not a theorem of set theory provable from the ZF axioms. Nor, as in the case of 8.1, does 8.4 represent a schema of theorems of set theory, each instance being a theorem of ZF. Strictly speaking, 8.4 is a theorem *about* ZF. It *will* guarantee that we *can* make recursive definitions *within* the ZF framework: but the principle *itself* is a theorem of formal logic. Expressed precisely, it says

---

†Readers who decided to skip this point in the context of 8.1 should do likewise here. For the casual reader, 8.4 as stated will cause no problems.

the following.

Suppose $\phi(v_o, v_1, v_2)$ is a formula of LAST such that

$$(\forall\alpha \in On)(\forall x)(\exists y)(\forall z)[z = y \leftrightarrow \phi(\alpha, x, z)].$$

Then there is a formula $\psi(v_o, v_1)$ of LAST such that the following are provable in ZF:

(a) $(\forall\alpha \in On)(\exists y)(\forall z)[z = y \leftrightarrow \psi(\alpha, z)]$

(b) $(\forall\alpha)(\forall y)[\psi(\alpha, y) \leftrightarrow (\exists z)(z \text{ is a function} \wedge \text{dom}(z) = \alpha$

$$\wedge (\forall\xi \in \alpha)\phi(\xi, z \upharpoonright \xi, z(\xi)) \wedge \phi(\alpha, z, y)]$$

We shall not give the proof in detail. In fact the idea is much as in 8.1, only now we cannot apply the replacement axiom as we did then to produce our function. Indeed, we cannot produce a function (working in ZF), since what we eventually get is a class. For the result to be at all relevant to our development of ZF, the only way to prove this is to actually *produce* a formula ψ as above. So what do we do? We take for our ψ precisely the LAST formula which appears on the right of the double arrow in (b) above. This makes (b) trivially true, and leaves us only to prove (a). (Actually, we must also check uniqueness, but this is really implicit in (b).)

We sketch the proof, using classes instead of formulas. Let $h : On \times V \to V$. Define a class f by

$$f = \{ (\alpha, x) \mid (\alpha \in On) \wedge (\exists z)[z \text{ is a function} \wedge \text{dom}(z) = \alpha$$

$$\wedge (\forall\xi \in \alpha)(z(\xi) = h(\xi, z \upharpoonright \xi)) \wedge x = h(\alpha, z)] \}.$$

It is easily seen that if $(\alpha, x), (\alpha, x') \in f$, then $x = x'$. And if there were an α such that no x existed with $(\alpha, x) \in f$, then consideration of the least such α would lead speedily to a contradiction. Hence $f : On \to V$. And clearly, $f(\alpha) = h(\alpha, f \upharpoonright \alpha)$ for all α. Finally, if $g : On \to V$ is such that $g(\alpha) = h(\alpha, g \upharpoonright \alpha)$, then by induction on α we get $f(\alpha) = g(\alpha)$ for all α, so $f = g$.

<u>Exercise 3</u>. *Fill in the details in the above sketch. Then give the proof without use of classes.*

9. <u>The Axiom of Choice</u>

There is one axiom of set theory which we have not discussed up to now : the Axiom of Choice. In its simplest form, this may be expressed as follows:

AC : Let $\mathfrak{F}$ be a set of pairwise disjoint, non-empty sets. Then there is a set M which consists of precisely one element from each member of $\mathfrak{F}$.

Now, in the case where $\mathfrak{F}$ is finite, the existence of such a set M is not problematical: we may prove it from the ZF axioms. But in general, when $\mathfrak{F}$ is infinite the existence of such an M cannot be proved in ZF. In certain specific cases it might be. For instance, suppose $\mathfrak{F}$ is a set of pairwise disjoint, non-empty sets *of ordinals*. Let $M = \{\alpha \in \cup\mathfrak{F} \mid (\exists X \in \mathfrak{F})(\alpha \text{ is the least member of } X)\}$. M is a well defined set by virtue of the axioms of union and subset selection. And M clearly consists of exactly one element from each member of $\mathfrak{F}$. The reason why we were able to construct such a set was that we had some rule for picking out (or *choosing*) one element of each set in $\mathfrak{F}$. In general, no such rule will be available to us. But even then, it is intuitively clear that such a set M will always "exist". We know that *all* subsets of $\cup\mathfrak{F}$ "exist": our set theory is totally unrestrictive. And does not this "all" include a set such as M? If it did not, would we not then have some (implicit) restriction on our ability to form sets? As we indicated earlier, there is *no* possibility of proving $\neg$AC from the ZF axioms. So, intuitively at least we can conclude that AC must be "true", and is implicit in our avowed freedom to form new sets from old ones. Having decided that AC is "true" (of our set theory), and knowing that it is not provable from the ZF axioms (which, let us inform you, we do), we therefore must adopt it as an axiom. Which is what we did when we formulated the Zermelo-Fraenkel system in sections 3 and 4.

Whilst AC as formulated above is the simplest version of the Axiom of Choice, it is by no means the most useful as far as applications are concerned. In this section we establish various alternative formulations. Now, normally we adopt the entire Zermelo-Fraenkel axioms (i.e. ZFC) as our basic set theory. But when we are proving that various forms of the Axiom of Choice are "equivalent", we clearly do not want to be using the Axiom of Choice itself as a fundamental *axiom* (i.e. something to be used freely). When we say that statement X is *equivalent* to AC, we mean, of course, that this equivalence can be established *in the system ZF alone!* To emphasise this point, we shall mark all the relevant theorems as being provable *in ZF*.

Our first reformulation of AC concerns *choice functions*. Let $\mathfrak{F}$ be a set of non-empty sets. A *choice function* for $\mathfrak{F}$ is a function $f : \mathfrak{F} \to \cup\mathfrak{F}$ such that for each $X \in \mathfrak{F}$, $f(X) \in X$. Let AC' be the assertion:

Every set of non-empty sets has a choice function.

9.1 <u>Theorem</u> (in ZF)

$AC \leftrightarrow AC'$.

Proof: $(\to)$ Let $\mathfrak{F}$ be a set of non-empty sets. For each $X \in \mathfrak{F}$, let $X^* = X \times \{X\}$. By the axiom of replacement, let $\mathfrak{F}^*$ be the set

$$\mathfrak{F}^* = \{ X^* \mid X \in \mathfrak{F} \} .$$

Clearly, $\mathfrak{F}^*$ is a set of non-empty, pairwise disjoint sets. By AC, let $M \subseteq \cup\mathfrak{F}^*$ be a set such that $M \cap X^*$ has exactly one element for each $X \in \mathfrak{F}$. Let $f^*(X)$ denote the unique element of $M \cap X^*$ for each $X \in \mathfrak{F}$. More formally, set

$$f^*(X) = \cup(M \cap X^*) .$$

Define $f : \mathfrak{F} \to \cup\mathfrak{F}$ by

$$f(X) = (f^*(X))_0$$

Clearly, f is a choice function for $\mathfrak{F}$.

(←) Let $\mathfrak{F}$ be a set of pairwise-disjoint, non-empty sets. By AC', let f be a choice function for $\mathfrak{F}$. Let $M = f[\mathfrak{F}]$. Clearly, M contains exactly one member of each X in $\mathfrak{F}$. □

Some authors actually regard AC' as the "basic" form of the Axiom of Choice. But as the above proof shows, AC and AC' are very close to each other, and after this chapter we shall not bother to distinguish between the two formulations, referring to both as AC and using the most convenient version in any instance.

Our next equivalence to AC is *Zermelo's Well Ordering Principle*. Let WO denote the assertion :

Every set can be well-ordered.

We shall prove that AC and WO are equivalent. First we need a lemma.

9.2 Lemma (in ZF)

Assume AC'. Let A be any set. Then there is a function $f : \mathcal{P}(A) \to A \cup \{A\}$ such that $f(A) = A$ and $f(X) \in A - X$ whenever $X \subset A$.

Proof : Let $B = \{ A - X \mid X \subset A \}$. By AC', let g be a choice function for B. Thus $g : B \to \cup B$ and $g(Y) \in Y$ for all $Y \in B$. Define $f : \mathcal{P}(A) \to A \cup \{A\}$ by

$$f(A) = A$$

$$f(X) = g(A - X) , \text{ if } X \subset A.$$

Clearly, f is as required. □

9.3 Theorem (in ZF)

$AC \leftrightarrow WO$.

Proof : (→) By the recursion principle, given a set A we may define a function $h : On \to V$ by

$$h(\alpha) = \begin{cases} f(\, h[\alpha] \cap A) \ , & \text{if } A \nsubseteq h[\alpha], \\ \{A\} \ , & \text{otherwise,} \end{cases}$$

where $f : \mathcal{P}(A) \to A$ is as in 9.2.

We claim that for some α, $h(\alpha) = \{A\}$. For suppose otherwise. Then $h(\alpha) \in A$ for all α. Hence by the axiom of subset selection,

$$X = h[On] = \{ a \in A \mid (\exists\alpha)(h(\alpha) = a) \}$$

is a set, and $h : On \to X$ is a surjection. In fact, h is a bijection. For if $\alpha < \beta$, then $h(\alpha) \in h[\beta] \cap A$, so as $f(\, h[\beta] \cap A \,)$ cannot lie in $h[\beta] \cap A$ (by choice of f), $h(\alpha) \neq h(\beta)$. Hence the inverse "function" $h^{-1} : X \to On$ exists. By the axiom of replacement, therefore, On is a set, contrary to Exercise 3 in section 6.

Now let α be least such that $h(\alpha) = \{A\}$. Thus

$$\gamma < \alpha \quad \to \quad h(\gamma) \in A.$$

But if $h[\alpha] \neq A$, then by definition of h, $h(\alpha) \neq \{A\}$. Hence

$$h : \alpha \leftrightarrow A.$$

We can thus well-order A by :

$$a <_A b \quad \leftrightarrow \quad h^{-1}(a) \in h^{-1}(b) \ .$$

($\leftarrow$) Let $\mathfrak{F}$ be a set of pairwise disjoint, non-empty sets. Let $X = \cup\mathfrak{F}$. By WO, let $<_X$ be a well-ordering of X. Let

$$M = \{ x \in X \mid (\exists A \in \mathfrak{F})(x \text{ is the } <_X\text{-least member of } A) \} .$$

Clearly, M satisfies AC for $\mathfrak{F}$. □

In conjunction with I.7.12, the above result yields at once the following corollary

9.4 Corollary (in ZF)

AC $\leftrightarrow$ For every set X there is an ordinal α and a bijection $f : \alpha \leftrightarrow X$. □

Our next equivalent to AC is perhaps the one most familiar to the working mathematician outside of set theory. For historical reasons it is known as a "lemma", but it is indeed just another formulation of the Axiom of Choice.

Let $(P, \leq)$ be a poset. An element a of P is said to be *maximal* in P iff there is no b in P such that $a < b$. A poset can have many maximal elements. The concept of a maximal element should not be confused with that of a *maximum* element. A *maximum* element of P is an element a of P such that $b \leq a$ for all b in P, and there can clearly be at most one such element. (In the case of tosets, the two concepts do, however, coincide, as is easily seen.)

A subset X of a poset $(P, \leq)$ is called a *chain* if it is totally ordered by $\leq$.

Zorn's Lemma (ZL) is the assertion : if a poset $(P, <_P)$ has the property that every chain in P has an upper bound in P, then P has a maximal element.

9.5 Theorem (in ZF)

AC $\rightarrow$ ZL.

Proof: Let $(P, \leq_P)$ be a poset such that every chain in P has an upper bound in P. By 9.4, let λ be an ordinal and let $j : \lambda \leftrightarrow P$. For each $\xi < \lambda$, let $p_\xi = j(\xi)$. Thus $P = \{p_\xi \mid \xi < \lambda\}$.

By the recursion principle, define $f : On \rightarrow \lambda+1$ so that $f(0) = 0$ and for $\eta > 0$,

$$f(\eta) = \begin{cases} \text{the least } \zeta \text{ such that } \xi < \eta \to p_{f(\xi)} <_P p_\zeta \text{ , if such a } \zeta \text{ exists,} \\ \\ \lambda \quad , \quad \text{otherwise.} \end{cases}$$

We claim that $f(\eta) = \lambda$ for some η. For suppose not. By the Axiom of Subset Selection, $X = f[On]$ is a well-defined subset of λ. Since f is one-one, f has a well-defined inverse, g, on X. Then $g : X \to On$ is surjective. But by the Axiom of Replacement, $g[X]$ is a set, so we have a contradiction.

Let η be least such that $f(\eta) = \lambda$. If η is a limit ordinal, then $\langle p_{f(\xi)} \mid \xi < \eta \rangle$ is a chain in P with no upper bound, which is impossible. Hence $\eta = \nu + 1$, for some ν. Clearly, $p_{f(\nu)}$ is a maximal element of P. □

The following variant of Zorn's lemma is also common, and will be denoted here by ZL'.

ZL' : If $(P, \leq_P)$ is a poset such that every chain in P has an upper bound in P, then for every $p \in P$ there is a $q \in P$, $p \leq_P q$, such that q is maximal in P.

9.6 <u>Theorem</u> (in ZF)

ZL $\to$ ZL'.

Proof : Let $(P, \leq_P)$ be as above. Let $p \in P$ be given. Set

$$Q = \{ q \in P \mid p \leq_P q \} .$$

With the induced ordering, Q is a poset which satisfies the hypotheses of ZL. By

ZL, let q be a maximal element in Q. Then $p \leq_P q$, and q is clearly maximal in P. □

Instead of proving directly that ZL' → AC now, we construct a chain of implications which will end up with AC, thereby establishing in one go a whole collection of equivalences to AC.

The *Hausdorff Maximal Principle* (HP) says that if $(P, \leq_P)$ is a poset, then every chain in P can be extended to a maximal chain.

9.7 Theorem (in ZF)

ZL' → HP.

Proof : Let $(P, \leq_P)$ be a given poset. Let $\mathfrak{F}$ be the set of all chains in P. $\mathfrak{F}$ is partially ordered by inclusion. We claim that the poset $(\mathfrak{F}, \subseteq)$ has the property that every chain in $\mathfrak{F}$ has an upper bound in $\mathfrak{F}$. In fact, if $\mathcal{C}$ is a chain in $\mathfrak{F}$, then $\cup\mathcal{C}$ is easily shown to be a member of $\mathfrak{F}$, and hence an upper bound of $\mathcal{C}$. Hence, applying ZL' to the poset $(\mathfrak{F}, \subseteq)$, we conclude that every member of $\mathfrak{F}$ extends to a maximal member of $\mathfrak{F}$. This is just HP. □

A set A is said to have *finite character* if $A \neq \emptyset$ and for any set X, X is a member of A iff every finite subset of X is a member of A.

Exercise 1. *Let A be any set. Let $\mathfrak{F}$ be the set of all subsets of $\mathcal{P}(A)$ which consist only of disjoint subsets of A. (i.e. if $X \in \mathfrak{F}$, then $X \subseteq \mathcal{P}(A)$ and $S, T \in X \to S \cap T = \emptyset$.) Show that $\mathfrak{F}$ is a set of finite character.*

Exercise 2. *Let A, B be any sets. Let $\mathfrak{F}$ be the set of all functions f such that $\mathrm{dom}(f) \subseteq A$ and $\mathrm{ran}(f) \subseteq B$. Show that if we regard $\mathfrak{F}$ as a subset of $\mathcal{P}(A \times B)$ (which, strictly speaking, it is), then $\mathfrak{F}$ is a set of finite character.*

Exercise 3. *Show that if we modify the example of Exercise 2 by insisting that* *f be one-one, then $\mathfrak{F}$ is still a set of finite character.*

Tukey's Lemma (TL) says that every set of finite character has an element which is maximal with respect to inclusion.

The concept of finite character is, at first sight, rather strange. The proof of the following result should indicate the type of circumstance under which TL can be applied.

9.8 Theorem (in ZF)

TL $\rightarrow$ AC'.

Proof : Let $\mathfrak{F}$ be a set of non-empty sets. We seek a function $f : \mathfrak{F} \rightarrow \cup\mathfrak{F}$ such that $f(X) \in X$ for every $X \in \mathfrak{F}$. Setting $A = \cup\mathfrak{F}$, it suffices to find an $f : \mathcal{P}(A) \rightarrow A$ such that $f(X) \in X$ for every non-empty $X \subseteq A$. Set

$$G = \{ f \mid f : \mathcal{P}(A) \rightarrow A \} .$$

For each $f \in G$, let

$$\mathfrak{F}(f) = \{ X \subseteq A \mid f(X) \in X \} .$$

Thus, f is a choice function for the family $\mathfrak{F}(f)$ of subsets of A. Set

$$K = \{ f \mid (\exists g \in G)(f \subseteq g \upharpoonright \mathfrak{F}(g)) \} .$$

K is a set of subsets of $\mathcal{P}(A) \times A$. It is easily seen that K has finite character. So, by TL, K has a maximal element, f_o. Suppose $\mathrm{dom}(f_o) \neq \mathcal{P}(A) - \{\emptyset\}$. Then we can find $X \subseteq A$, $X \notin \mathrm{dom}(f_o)$, $X \neq \emptyset$. Pick $x \in X$ arbitrarily, and set $f_o' = f_o \cup \{ (X, x) \}$. Then $f_o' \in K$ and $f_o \subset f_o'$, contrary to the choice of f_o. Hence $\mathrm{dom}(f_o) = \mathcal{P}(A) - \{\emptyset\}$. Thus f will be as required, where we set $f = f_o \cup \{ (\emptyset, a) \}$, with a any element of A. □

The next result completes our chain of implications proving that AC, ZL, HP and TL are equivalent.

9.9 Theorem. (in ZF)

HP $\to$ TL.

Proof : Let $\mathfrak{F}$ be a set of finite character. Regarding $\mathfrak{F}$ as a poset under inclusion, let $\mathcal{C}$ be a maximal chain in $\mathfrak{F}$ (by HP). Now, if $\mathcal{C}$ were to have a greatest element, then the maximality of $\mathcal{C}$ would mean that such an element would be maximal in $\mathfrak{F}$, and so we would be done. We show that $\mathcal{C}$ does in fact have a greatest element. Suppose otherwise. Set $A = \cup\mathcal{C}$. Since $\mathcal{C}$ has no greatest element, $A \not\in \mathcal{C}$. Hence we have $X \in \mathcal{C} \to X \subset A$. Now, if $A \in \mathfrak{F}$, $\mathcal{C} \cup \{A\}$ would be a chain in $\mathfrak{F}$ extending $\mathcal{C}$. Hence $A \not\in \mathfrak{F}$. Thus as $\mathfrak{F}$ has finite character, there is a finite set $a \subseteq A$ such that $a \not\in \mathfrak{F}$. Since $a \subseteq A = \cup\mathcal{C}$ is finite and $\mathcal{C}$ has no last member, there is an $X \in \mathcal{C}$ such that $a \subseteq X$. But $X \in \mathfrak{F}$ and $\mathfrak{F}$ has finite character. Hence $a \in \mathfrak{F}$, a contradiction. □

Problems

1. (ϵ - Induction; ϵ - Recursion) More general than the notions of induction and recursion on ordinals are the notions of ϵ-induction and recursion on ϵ.
(A) If A is a class of sets such that for every x, $(\forall y \in x)(y \in A) \to (x \in A)$ (i.e. $x \subseteq A \to x \in A$), then $A = V$. (Principle of Proof by ϵ-Induction.)
(B) Let $h : V \times V \to V$. Then there exists a unique $f : V \to V$ such that for every set x, $f(x) = h(x, f \upharpoonright x)$. (Principle of ϵ-Recursion.)
(C) Define the function $\rho = V \to On$ by the ϵ-recursion

$$\rho(x) = \cup\{ \rho(y)+1 \mid y \in x \} .$$

$\rho(x)$ is the *rank* of x. $\rho(x)$ is the least γ such that $x \in V_{\gamma+1}$.

2. (Ideals and Filters) For basic definitions see Problems I.2. Let B be a boolean algebra, I an ideal in B, F a filter in B. I (resp. F) is *prime* iff for each b in B, either $b \in I$ or $-b \in I$ (resp. F).

(A) I (resp F) is prime iff it is maximal (i.e. is not B and is not contained in any ideal (resp. filter) other than B itself.)

(B) Let $\mathfrak{F}$ be a field of subsets of a set X. Let $x \in X$. The set of all sets A in $\mathfrak{F}$ with $x \notin A$ is a maximal ideal in $\mathfrak{F}$. The set of all sets A in $\mathfrak{F}$ with $x \in A$ is a maximal filter in $\mathfrak{F}$.

(C) Let X be an infinite set, and let $\mathfrak{F}$ be the set of all sets $A \subseteq X$ such that either A or X - A is finite. $\mathfrak{F}$ is a field of sets. The set of all finite (resp. infinite) sets in $\mathfrak{F}$ is a maximal ideal (resp. filter) in $\mathfrak{F}$.

(D) There is a natural one-one correspondence between maximal ideals, maximal filters, and boolean morphisms into the two element algebra $\underset{\sim}{2} = \{0, 1\}$.

(E) Every ideal other than B can be extended to a maximal ideal. (Uses AC). Similarly for filters.

Prime filters are often referred to as *ultrafilters*.

3. (Use of AC) The following results make essential use of the axiom of choice. (In some cases it requires careful thought to spot the usage.)

(A) The union of a countable set of countable sets is countable.

(B) Any vector space has a basis.

(C) There is a set of real numbers that is not Lebesgue measurable.

(D) A product of compact topological spaces is compact. (Tychonoff Theorem)

(E) In a Banach space $\underset{\sim}{B}$, any bounded linear functional defined on a subspace of $\underset{\sim}{B}$ extends to a bounded linear functional of the same norm defined on all of $\underset{\sim}{B}$. (Hahn-Banach Theorem)

(F) Any subgroup of a free abelian group is free abelian. (Nielsen-Schreier Theorem)

(G) Every boolean algebra is isomorphic to a field of sets. (Stone's Theorem)

Chapter III. Ordinal and Cardinal Numbers

1. Ordinal Numbers

The concept of an *ordinal number* (or *ordinal*) was introduced in I.7. We defined an ordinal to be a woset $(X, <)$ such that

$$a = \{ x \in X \mid x < a \}$$

for every $a \in X$. We proved that any two ordinals are either identical or else non-isomorphic, and moreover that if X, Y are non-identical ordinals, then either $X \in Y$ or else $Y \in X$. (Also, if $(X, <)$ is an ordinal, then the ordering $<$ is just $\subset$ on X is just $\in$ on X, which justifies our referring simply to X, Y above.) The first ordinal is 0, the second is $1 = \{0\}$, the n'th is $n-1 = \{0, \ldots, n-2\}$, the first infinite ordinal is $\omega = \{0, 1, 2, \ldots, n, n+1, \ldots\}$, the second infinite ordinal is $\omega+1 = \{0, 1, 2, \ldots, n, \ldots, \omega\}$, and so on; in general, the first ordinal after α is $\alpha+1 = \alpha \cup \{\alpha\}$. Any ordinal of the form $\gamma = \alpha+1$ (i.e. $\gamma = \alpha \cup \{\alpha\}$) is called a *successor ordinal*, and we write $\mathrm{succ}(\gamma)$. An ordinal, δ, not of this form is called a *limit ordinal*, and we write $\lim(\delta)$. And we have now adopted once more our convention that lower case Greek letters denote ordinals (with ω the only one with a specific meaning).

The following set theoretic characterisation of ordinals is very useful. Let us call a set X *transitive* iff

$$x \in X \ \& \ a \in x \ \rightarrow \ a \in X .$$

1.1 Lemma

A set X is[†] an ordinal iff it is transitive and totally ordered by ϵ.

Proof : Suppose X is an ordinal. That is, $(X, \subset)$ is a woset and for every $x \in X$, $x = \{a \in X \mid a \subset x\}$. Since $x \in X \rightarrow x \subseteq X$, X is transitive. And we know that X is totally ordered by ϵ.

Conversely, let X be a transitive set which is totally ordered by ϵ. By the axiom of foundation, X is thus well-ordered by ϵ. Now let $x \in X$. Since X is transitive, $a \in x \rightarrow a \in X$, so $x = \{a \in X \mid a \in x\}$. Thus X is an ordinal. □

Using 1.1, we can now *prove*, using the ZF axioms, that there are many ordinals.

By the null set axiom, the ordinal 0 exists. The existence of all the finite ordinals now follows from the following Lemma:

1.2 Lemma

If α is an ordinal, then $\alpha \cup \{\alpha\}$ is an ordinal.

Proof : If α is transitive and totally ordered by ϵ, so too is $\alpha \cup \{\alpha\}$. □

Instrumental in proving that there are many limit ordinals is the next lemma.

1.3 Lemma

If A is a set of ordinals, then $\cup A$ is an ordinal.

Proof : Let $x \in a \in \cup A$. For some $b \in A$, $a \in b$. Since b is an ordinal, $x \in a \in b$ implies $x \in b$. Hence $x \in \cup A$. Thus A is transitive.

Again, let $x, y \in \cup A$. Pick $a, b \in A$ with $x \in a$, $y \in b$. Either $a \subseteq b$

[†]It should be understood here that we mean X with the ordering $\subset$.

or $b \subseteq a$. Assume for the sake of argument that $a \subseteq b$. Then $x, y \in b$. Hence either $x \in y$ or else $y \in x$ (or else $x = y$). Hence $\cup A$ is totally ordered by $\in$. Thus $\cup A$ is an ordinal. □

Having observed that the ZF axioms guarantee the existence of all the finite ordinals, the next step is to obtain ω. Now, the existence of the ordinal ω follows from the axiom of infinity (together with some other axioms). But the construction of the set ω presents some technical difficulties, so instead of giving the proof here, we shall leave it as an exercise for the reader. (It is not hard, but does require some thought.) Given ω, the existence of the ordinals $\omega+1$, $\omega+2$ $(= (\omega+1) + 1)$, etc. now follows using 1.2 as before. Now let $\omega + \omega$ denote the next limit ordinal, i.e. the "set" $\{0, 1, 2, \ldots, \omega, \omega+1, \ldots\}$. That this set really "exists" (i.e. can be formed using the ZF axioms) may be demonstrated thus. Let the "function" $f : \omega \to V$ be defined by $f(n) = \omega+n$. By the Axiom of Replacement, $E = \{f(n) \mid n \in \omega\}$ is a set. Let $A = \cup E$. By the Axiom of Union, A is a set. By 1.3, A is an ordinal. Clearly, A is our ordinal $\omega + \omega$. And so on.

2. Addition of Ordinals

Given ordinals α, β, we define the ordinal $\alpha + \beta$. Intuitively, $\alpha + \beta$ is the ordinal which "commences" with α and continues beyond α for β more steps. (i.e. $\alpha + \beta$ is α "followed by" β.) Formally, we set

$$A = (\alpha \times \{0\}) \cup (\beta \times \{1\}),$$

and we define a well-ordering of A by

$$\langle \nu, i\rangle <_A \langle \tau, j\rangle \quad \text{iff} \quad (i < j) \vee (i = j \wedge \nu < \tau).$$

We then set : $\alpha + \beta = \mathrm{Ord}(\langle A, <_A\rangle)$.

It is immediate that the sum $\alpha + 1$ is the successor to α, so our previous notation for successor ordinals causes no problems. Moreover, $\alpha + n$ is the n'th ordinal beyond α, and $\alpha + \beta$ the β'th ordinal beyond α for any β.

2.1 <u>Lemma</u>

Ordinal addition is associative : for all α, β, γ,

$$\alpha + (\beta + \gamma) = (\alpha + \beta) + \gamma.$$

Proof : An easy exercise. □

Notice that ordinal addition is not commutative. For example, the following are easily verified :

$$1 + \omega = \omega \quad \text{but} \quad \omega + 1 > \omega.$$

Indeed, for any integer n, we have $n + \omega = \omega$, whereas

$$\omega < \omega + 1 < \omega + 2 < \omega + 3 < \ldots .$$

Using the addition concept, we can now present a fuller "picture" of the ordinal number system.

$0, 1, 2, \ldots, n, \ldots, \omega, \omega + 1, \omega + 2, \ldots, \omega + n, \ldots$

$\ldots, \omega + \omega, \omega + \omega + 1, \omega + \omega + 2, \ldots, \omega + \omega + n, \ldots, \omega + \omega + \omega,$

$\omega + \omega + \omega + 1, \omega + \omega + \omega + 2, \ldots .$

3. <u>Multiplication of Ordinals</u>

Let λ be an ordinal, and let $\langle \alpha_\xi \mid \xi < \lambda \rangle$ be a λ-sequence of ordinals. The ordinal sum

$$\sum_{\xi<\lambda} \alpha_\xi$$

is defined as follows.

Set $$A = \bigcup_{\xi<\lambda} (\alpha_\xi \times \{\xi\}).$$

Define a well-ordering of A by

$$\langle \nu, \xi \rangle <_A \langle \nu', \xi' \rangle \leftrightarrow (\xi < \xi') \vee (\xi = \xi' \wedge \nu < \nu').$$

Let
$$\sum_{\xi<\lambda} \alpha_\xi = \mathrm{Ord}(\langle A, <_A \rangle).$$

Clearly, the intuitive "picture" of $\sum_{\xi<\lambda} \alpha_\xi$ is the ordinal which commences with α_o, then has α_1 more steps, then another α_2 steps, and so on, up to all $\xi < \lambda$. For instance, we have

$$\sum_{\xi<2} \alpha_\xi = \alpha_o + \alpha_1 ,$$

$$\sum_{\xi<3} \alpha_\xi = \alpha_o + \alpha_1 + \alpha_2 ,$$

$$\sum_{\xi<n} \alpha_\xi = \alpha_o + \alpha_1 + \alpha_2 + \ldots + \alpha_{n-1}.$$

Notice that, in particular, $\sum_{n<\omega} n = \sum_{n<\omega} 1 = \omega.$

We may now define ordinal multiplication as iterated addition. That is, we define
$$\alpha.\beta = \sum_{\xi<\beta} \alpha .$$

Thus, $\alpha.\beta$ denotes "β copies of α", or "α followed by α followed by α ... (β times)". In particular, for any finite ordinal n,

$$\alpha.n = \underbrace{\alpha + \alpha + \ldots + \alpha}_{n \text{ times}}$$

The first thing to notice about ordinal multiplication is that it is not commutative. For instance, we clearly have

$$2.\omega = \omega \quad \text{but} \quad \omega.2 = \omega + \omega > \omega.$$

Indeed, for any finite ordinal n, $n.\omega = \omega$, but

$$\omega < \omega.2 < \omega.3 < \omega.4 < \ldots .$$

We do have a distributive law, namely:

3.1 <u>Lemma</u>

For any α, β, γ : $\alpha.(\beta + \gamma) = \alpha.\beta + \alpha.\gamma$

Proof : An easy exercise. □

The other distributivity property is false. For example

$$(1 + 1).\omega = 2.\omega = \omega$$

$$1.\omega + 1.\omega = \omega + \omega > \omega.$$

Finally, we have associativity of ordinal multiplication.

3.2 <u>Lemma</u>

For any α, β, γ : $(\alpha.\beta).\gamma = \alpha.(\beta.\gamma)$.

Proof : A moderately easy exercise. □

Using ordinal multiplication, we may now describe the ordinal number system even more fully than before:

$0, 1, 2, \ldots, n, \ldots, \omega, \omega + 1, \omega + 2, \ldots, \omega + n, \ldots, \omega + \omega,$
$\omega + \omega + 1, \omega + \omega + 2, \ldots, \omega + \omega + n, \ldots, \omega.3, \omega.3 + 1, \omega.3 + 2,$
$\ldots, \omega.3 + n, \ldots, \omega.4, \omega.4 + 1, \ldots, \omega.5, \ldots, \omega.n, \ldots, \omega.\omega,$
$\omega.\omega{+}1, \ldots, \omega.\omega.2, \ldots, \omega.\omega.n, \ldots, \omega.\omega.\omega, \ldots$

Notice that the limit ordinals are just those ordinals of the form $\omega.\alpha$ for an ordinal α. This indicates that the ordinals consist of nothing more than an "endless" sequence of copies of ω placed one after the other. However, although this is strictly speaking true, it provides the beginner with a picture which is almost certainly false. The deep implications which lie behind the word "endless" mean that there are many limit ordinals which do not resemble ω in the least (even though they *are* of the form $\omega.\alpha$ for some α!). This will become clear when we are able to describe some ordinals which are much bigger than any mentioned above.

The remaining basic arithmetical operation is exponentiation. But before we can introduce this notion, we need to establish some fundamental results about sequences of ordinals.

4. Sequences of Ordinals

Let λ be a limit ordinal, and let $\langle\alpha_\xi \mid \xi < \lambda\rangle$ be a λ-sequence of ordinals. We write

$$\alpha = \lim_{\xi<\lambda} \alpha_\xi$$

iff $(\forall\beta < \alpha)(\exists\xi < \lambda)(\forall\zeta)(\xi < \zeta < \lambda \rightarrow \beta < \alpha_\zeta \leq \alpha)$.

We then call α the *limit* of the sequence $\langle\alpha_\xi \mid \xi < \lambda\rangle$. (If such an α exists, it is clearly unique.)

Our next lemma shows that many sequences do have limits.

4.1 Lemma

Let $\lim(\lambda)$, and let $\langle\alpha_\xi \mid \xi < \lambda\rangle$ be a monotone increasing sequence of ordinals. Then $\langle\alpha_\xi \mid \xi < \lambda\rangle$ has a (unique) limit; and indeed,

$$\lim_{\xi<\lambda} \alpha_\xi = \bigcup_{\xi<\lambda} \alpha_\xi .$$

Proof : An easy exercise. □

4.2 Lemma

Let λ, μ be limit ordinals, and let $f : \mu \rightarrow \lambda$ be an order preserving function such that $\lim_{\xi<\mu} f(\xi) = \lambda$. Let $\langle\alpha_\xi \mid \xi < \lambda\rangle$ be an increasing sequence. Then

$$\lim_{\xi<\lambda} \alpha_\xi = \lim_{\xi<\mu} \alpha_{f(\xi)} .$$

Proof : An easy exercise. □

4.3 Lemma

Let lim(λ), and let $\langle \alpha_\xi \mid \xi < \lambda \rangle$, $\langle \beta_\zeta \mid \zeta < \lambda \rangle$ be increasing sequences such that

(a) $(\forall \xi < \lambda)(\exists \zeta < \lambda)(\beta_\zeta > \alpha_\xi)$;

(b) $(\forall \zeta < \lambda)(\exists \xi < \lambda)(\alpha_\xi > \beta_\zeta)$.

Then

$$\lim_{\xi<\lambda} \alpha_\xi = \lim_{\zeta<\lambda} \beta_\zeta .$$

Proof : An easy exercise. □

4.4 Lemma

Let lim(λ), and let $\langle \alpha_\xi \mid \xi < \lambda \rangle$ be any λ-sequence of ordinals. For each $\mu < \lambda$, let

$$\sigma_\mu = \sum_{\xi<\mu} \alpha_\xi .$$

Then

$$\sum_{\xi<\lambda} \alpha_\xi = \lim_{\mu<\lambda} \sigma_\mu .$$

Proof: We leave the proof as an exercise. □

Let $f : \lambda \to \lambda$, and let $\alpha \in \lambda$ be a limit ordinal. We say f is *continuous at* α iff

$$f(\alpha) = \lim_{\xi<\alpha} f(\xi).$$

For example, the identity function on λ is continuous at every limit ordinal in λ. (We shall see many more examples of continuity later.)

Exercise 1. *Let λ be endowed with the order topology (see Problem I.3). Show that a function $f : \lambda \to \lambda$ is continuous at α in the sense just defined iff it is continuous at α with respect to the topology on λ.*

A function $f : \lambda \to \lambda$ is a *normal function* iff it is order preserving and continuous at every limit ordinal in λ.

4.5 Lemma

Let $f : \mu \to \mu$ be a normal function, and let $\lambda \in \mu$ be a limit ordinal. If $\langle \alpha_\xi \mid \xi < \lambda \rangle$ is an increasing sequence of ordinals in μ and $\lim_{\xi<\lambda} \alpha_\xi < \mu$, then

$$f(\lim_{\xi<\lambda} \alpha_\xi) = \lim_{\xi<\lambda} f(\alpha_\xi) .$$

Proof : An easy exercise. □

Concerning normal functions, the following lemma is often useful.

4.6 Lemma

Let $f : \lambda \to \lambda$ be order preserving. Then $f(\alpha) \geq \alpha$ for all $\alpha \in \lambda$.

Proof : By induction on α. For $\alpha = 0$ there is nothing to prove. Assuming $f(\alpha) \geq \alpha$, then $f(\alpha + 1) > f(\alpha) \geq \alpha$, so $f(\alpha + 1) \geq \alpha + 1$. Finally, if $\lim(\alpha)$ and $f(\beta) \geq \beta$ for all $\beta < \alpha$, then since $f(\alpha) > f(\beta)$ for all $\beta < \alpha$, we have $f(\alpha) > \beta$ for all $\beta < \alpha$, so $f(\alpha) \geq \bigcup_{\beta<\alpha} \beta = \alpha$. □

Let $f : \lambda \to \lambda$. We say $\alpha \in \lambda$ is a *fixed-point* of f iff $f(\alpha) = \alpha$.

4.7 Theorem. (Fixed-Point Theorem for Normal Functions on On.)

Let $f : On \to On$ be a normal function (in the class sense). For every α there is a fixed-point γ of f such that $\gamma \geq \alpha$.

Proof: Let α be given. If $f(\alpha) = \alpha$ there is nothing further to prove. So assume otherwise. Then, by 4.6, $f(\alpha) > \alpha$. By recursion, we define a function $g : \omega \to On$ so that

$$g(0) = \alpha$$

$$g(n+1) = f(g(n)).$$

An easy induction proves that g is order-preserving. By 4.1, let $\gamma = \lim_{n<\omega} g(n)$. Notice that $\gamma > g(0) = \alpha$. We finish by proving that $f(\gamma) = \gamma$. Since f is a normal function, we have, by 4.5

$$f(\gamma) = f(\lim_{n<\omega} g(n)) = \lim_{n<\omega} f(g(n)) = \lim_{n<\omega} g(n+1) = \gamma. \quad \square$$

The above result does not hold if $f : \lambda \to \lambda$ (in general). For instance, the function $f : \omega \to \omega$ defined by $f(n) = n + 1$ has no fixed points. There do exist ordinals λ such that every normal function $f : \lambda \to \lambda$ has a fixed point, and indeed arbitrarily large fixed-points, but we shall not be able to characterise these ordinals until later.

5. Ordinal Exponentiation

Let $\alpha \in On$. By recursion, we define a function $f_\alpha : On \to On$ so that :

$$f_\alpha(0) = 1 ;$$

$$f_\alpha(\beta + 1) = f_\alpha(\beta).\alpha ;$$

$$f_\alpha(\beta) = \lim_{\gamma<\beta} f_\alpha(\gamma) \text{ , if } \lim(\beta) .$$

We write α^β instead of $f_\alpha(\beta)$. Thus, α^β is defined by "the recursion":

$$\alpha^0 = 1 ;$$

$$\alpha^{\beta+1} = \alpha^\beta.\alpha ;$$

$$\alpha^\beta = \lim_{\gamma<\beta} \alpha^\gamma , \text{ if } \lim(\beta) .$$

Thus, α^β corresponds to the product of α with itself taken β times. In particular, $\alpha^1 = \alpha$, $\alpha^2 = \alpha.\alpha$, $\alpha^3 = \alpha.\alpha.\alpha$, ...

5.1 Lemma

(i) $\alpha^\beta.\alpha^\gamma = \alpha^{(\beta+\gamma)}$.

(ii) $(\alpha^\beta)^\gamma = \alpha^{(\beta.\gamma)}$.

Proof : In each case fix α and β and argue by induction on γ. The details are left as an exercise. □

5.2 Lemma

Let α be fixed. Regarded as functions of β, the functions $\alpha + \beta$, $\alpha.\beta$, α^β are normal functions.

Proof : Exercise. □

5.3 Corollary

For any α, there are arbitrarily large ordinals β, γ, δ such that

$$\alpha + \beta = \beta , \quad \alpha.\gamma = \gamma , \quad \alpha^\delta = \delta .$$

Proof : By 4.7. □

<u>Exercise 1.</u> *Show that for any* α, $\alpha + \alpha.\omega = \alpha.\omega$ *and* $\alpha.\alpha^{\omega} = \alpha^{\omega}$.

<u>Exercise 2.</u> *Show that for any finite* $n > 1$, $n^{\omega} = \omega$.

Let us now use the notion of ordinal exponentiation in order to extend our picture of the ordinal number system somewhat.

$$0, 1, 2, \ldots, \omega, \omega + 1, \ldots, \omega.2, \ldots, \omega.3, \ldots, \omega.\omega, \ldots$$
$$\ldots, \omega^3, \omega^3 + 1, \ldots, \omega^3 + \omega, \ldots, \omega^3 + \omega^2, \ldots, \omega^4, \omega^4 + 1, \ldots,$$
$$\ldots, \omega^{\omega}, \omega^{\omega} + 1, \ldots, \omega^{\omega.2}, \ldots, \omega^{\omega.3}, \ldots, \omega^{\omega.\omega}, \ldots,$$
$$\omega^{(\omega^2)}, \ldots, \omega^{(\omega^3)}, \ldots, \omega^{(\omega^{\omega})}, \ldots \quad \text{(and so on).}$$

We are thus able to picture very many ordinal numbers. Nevertheless, as we shall see in the remaining parts of this chapter, the above picture does not even begin to describe the true situation. The above "sequence" is only an "infinitesimal" initial part of the sequence of all ordinal numbers. Even the "giant" ordinal

$$\left.\omega^{\omega^{\omega^{\cdot^{\cdot^{\cdot}}}}}\right\} \omega \text{ times}$$

is tiny in comparison with "most" ordinal numbers.

6. <u>Cardinality. Cardinal Numbers.</u>

We are now in a position to assign to every set a quantity which represents the "size" or "number of elements" of that set. In the case of a finite set, our notion really will be the number of elements of the set. For infinite sets we shall obtain a generalisation of this finite concept.

Let us commence by considering finite sets. If A is a finite set, let n(A) denote the number of elements of A. In essence n(A) is some sort of abstraction from A with the property that if A and B are two finite sets, then

(I) $n(A) = n(B)$ iff A and B can be put into one-one correspondence.

What exactly *is* the object $n(A)$? It is a finite ordinal. The ordinal m is a set with exactly m elements. Thus :

(II) $n(m) = m$.

By (I) and (II), we have (for A still finite)

(III) $n(A) = m$ iff A and m can be put into one-one correspondence.

We turn now to the general case. By II.9.4 (which uses AC), if X is any set, there is an ordinal α and a bijection $f : \alpha \leftrightarrow X$. Which might suggest that we can extend our previous notion of "number of elements" from the finite to the infinite realm by just using the ordinals. Unfortunately, this does not work. If X is infinite, there is not a unique α as above, but infinitely many such. For example, the set $\omega = \{0, 1, 2, \dots\}$ can be put into one-one correspondence with the ordinal ω by means of the identity map, and with the ordinal $\omega.2$ by means of the bijection

$$f(n) = \begin{cases} n/2, & \text{if } n \text{ is even,} \\ \omega + \frac{n-1}{2}, & \text{if } n \text{ is odd.} \end{cases}$$

Thus, although the finite ordinals provide us with an excellent number system for gauging the size of finite sets, the same cannot be said of the infinite ordinals for infinite sets. At least, not if we try to do it in a naive manner. But if we make use of the fact that the ordinals are well-ordered by $\in$, we can easily obtain a suitable number system for "measuring" arbitrary sets.

By II.9.4, we know that for any set X there is an ordinal α and a bijection $f : \alpha \leftrightarrow X$. The *cardinality* of X, denoted by $|X|$, is the *least* ordinal α for which there exists a bijection $f : \alpha \leftrightarrow X$. $|X|$ is uniquely defined, and will represent the "number of elements" of X. Clearly, if X is finite, then $|X| = n(X)$ as defined earlier. Moreover, analogues of properties (I), (II) and

(III) above hold in the generalised situation. (This is clear from the definition.)

Of course, although we are using the ordinal number system to "measure" our sets, we are not using all the ordinal numbers. For instance, our remark above shows that the ordinal $\omega.2$ is never the cardinality of a set.

A *cardinal number* (or *cardinal*) is an ordinal, α, such that for no $\beta < \alpha$ does there exist a bijection $f : \beta \leftrightarrow \alpha$. It is immediate that the cardinality of any set is in fact a cardinal number, and, conversely, any cardinal number is the cardinality of some set. (In fact, the cardinal number α is the cardinality of the set $\alpha = \{\beta \mid \beta < \alpha\}$.) It is customary to restrict the letters κ, λ, μ to denote cardinals, though λ and μ are sometimes used to denote arbitrary limit ordinals.

6.1 <u>Lemma</u>

(i) Every finite ordinal is a cardinal.

(ii) ω is a cardinal.

(iii) Every infinite cardinal is a limit ordinal.

Proof : (i) and (ii) are immediate. We prove (iii). Let $\alpha \geq \omega$. We show that $\alpha + 1$ is not a cardinal. Define $f : \alpha \to \alpha + 1$ by

$$f(0) = \alpha$$
$$f(n+1) = n$$
$$f(\xi) = \xi, \quad \text{if } \omega \leq \xi < \alpha.$$

Clearly, f is a bijection. Hence $\alpha + 1$ is not a cardinal. □

Now, the notions of cardinality and of cardinal number were defined using bijections. But it is often quite tricky to construct a bijection to verify some assertion about cardinality or cardinal numbers. In such instances, the theorem proved below is often helpful. We need a simple lemma.

6.2 Lemma

Let X, Y be sets. Then $|X| \leqslant |Y|$ iff there is an injection $f : X \to Y$.

Proof: Let $\kappa = |X|$, $\lambda = |Y|$, and let $i : \kappa \leftrightarrow X$, $j : \lambda \leftrightarrow Y$.

Suppose first that there is an injection $f : X \to Y$. Let $h = j^{-1} \circ f \circ i$. Then $h : \kappa \to \lambda$ is an injection. Let $U = h[k]$. Since $U \subseteq \lambda$, U is well-ordered (by the ordinal relation <). Let $\gamma = \mathrm{Ord}(U)$ (see page 28), and let $\pi : \gamma \leftrightarrow U$.

By definition of cardinality, $|U| \leqslant \gamma$. But clearly, $\gamma \leqslant \lambda$. Hence $|U| \leqslant \lambda$. Since $h : \kappa \leftrightarrow U$, $|\kappa| = |U|$, and it follows that $|\kappa| \leqslant \lambda$, i.e. $\kappa \leqslant \lambda$.

Now suppose $\kappa \leqslant \lambda$. Then $j \circ i^{-1} : X \to Y$ is a well-defined injection. □

6.3 Theorem (Schröder-Bernstein)

Let X, Y be sets. If there are injections $i : X \to Y$ and $j : Y \to X$, then there is a bijection $f : X \leftrightarrow Y$.

Proof: Let $\kappa = |X|$, $\lambda = |Y|$ and let $h : \kappa \leftrightarrow X$, $k : \lambda \leftrightarrow Y$. By 6.2, $\kappa \leqslant \lambda$ and $\lambda \leqslant \kappa$. Hence $\kappa = \lambda$. Let $f = k \circ h^{-1}$. □

Exercise 1 : *The above theorem was proved with the aid of the Axiom of Choice, using the notion of cardinality. The "classical" proof of the result, though a little more complicated, proceeds by a direct combinatorial argument which does not use the Axiom of Choice. The proof is outlined below. Your task is to fill in the details.*

1. *The first step is to show that if* X *is any set and* $h: \mathcal{P}(X) \to \mathcal{P}(X)$ *is such that*

$$A \subseteq B \subseteq X \to h(A) \subseteq h(B)$$

then there is a set $T \subseteq X$ *such that* $h(T) = T$.

(Hint: Set $T = \bigcup\{A \subseteq X \mid A \subseteq h(A)\}$.*)*

2. *Given sets* X, Y *and injections* $i : X \to Y$, $j : Y \to X$ *now, define a function* $* : \mathcal{P}(X) \to \mathcal{P}(X)$ *by setting*

$$A^* = X - j[Y - i[A]]$$

for each $A \subseteq X$.

Then $A \subseteq B \subseteq X \to A^* \subseteq B^*$.

3. *Combining parts 1 and 2 above, there is a set* $T \subseteq X$ *such that* $T^* = T$, *i.e.*

$$T = X - j[Y - i[T]] .$$

Define $f : X \to Y$ *by*

$$f(x) = \begin{cases} i(x), & \text{if } x \in T \\ j^{-1}(x), & \text{if } x \in X - T . \end{cases}$$

Then f is a bijection, as required.

Using the Schröder-Bernstein Theorem we can easily obtain an alternative characterisation of cardinal numbers. First a simple lemma.

6.4 Lemma

Let X,Y be non-empty sets. The following are equivalent:

(i) There is an injection $f : X \to Y$.

(ii) There is a surjection $g : Y \to X$.

Proof: (i) $\to$ (ii) Choose $x_o \in X$ arbitrarily. Define $g : Y \to X$ by

$$g(y) = \begin{cases} x, & \text{if } x \text{ is the unique member of } X \text{ such that } f(x) = y \\ x_o, & \text{if there is no } x \in X \text{ such that } f(x) = y. \end{cases}$$

Clearly, g is a surjection.

(ii) $\to$ (i). Let $<_Y$ be a well-ordering of Y. Define $f : X \to Y$ by setting

$$f(x) = \text{the } <_Y\text{-least } y \in Y \text{ such that } g(y) = x.$$

Clearly, f is an injection. □

6.5 Lemma

An ordinal α is a cardinal iff for no ordinal $\beta < \alpha$ is there a surjection $f : \beta \to \alpha$.

Proof: If α is not a cardinal, there is a $\beta < \alpha$ and a *bijection* $f : \beta \to \alpha$, so we are done.

Now suppose that there is a $\beta < \alpha$ and a surjection $f : \beta \to \alpha$. By 6.4 there is thus an injection $g : \alpha \to \beta$. But $\beta < \alpha$, so $id_\beta : \beta \to \alpha$ is an injection. By the Schröder-Bernstein theorem there is thus a bijection $h : \beta \leftrightarrow \alpha$. Hence α cannot be a cardinal. □

So far we have only met one infinite cardinal, ω. Our next result shows that there are at least infinitely many infinite cardinals.

6.6 Lemma

If κ is a cardinal, there is a cardinal greater than κ.

Proof : Let $X = \mathcal{P}(\kappa)$, $\lambda = |X|$. We show that $\lambda > \kappa$. Well, since the map $j : \kappa \to X$ defined by $j(\alpha) = \{\alpha\}$ is an injection, 6.2 tells us that $\lambda \geq \kappa$. Suppose $\lambda = \kappa$. Thus there is a bijection $f : \kappa \leftrightarrow X$. Let

$$A = \{\alpha \in \kappa \mid \alpha \notin f(\alpha)\}.$$

Clearly, A is a well-defined subset of κ. So, for some $\alpha_o \in \kappa$, we must have $A = f(\alpha_o)$. Then,

$$\alpha_o \in A \leftrightarrow \alpha_o \notin f(\alpha_o) \leftrightarrow \alpha_o \notin A.$$

This contradiction completes the proof. □

Of course, since the ordinals are well-ordered by $\in$, so too are the cardinals. Hence, for each cardinal κ there is a unique least cardinal greater than κ : this cardinal is denoted by κ^+, and referred to as the *successor cardinal* to κ (or simply the *successor* to κ, when there is no possible confusion with the successor *ordinal*, $\kappa + 1$).

The first cardinal after ω is denoted by ω_1; the next cardinal by ω_2, and so on, providing an infinite sequence of infinite cardinals

$$\omega, \omega_1, \omega_2, \ldots, \omega_n, \omega_{n+1}, \ldots$$

Our next result shows that the sequence does not stop after ω steps.

6.7 Lemma

Let $\lim(\delta)$, and let $\langle\kappa_\xi \mid \xi < \delta\rangle$ be a strictly increasing sequence of infinite cardinals. Let $\kappa = \lim_{\xi<\delta} \kappa_\xi$. Then κ is a cardinal.

Proof : By 4.1, $\kappa = \bigcup_{\xi<\delta} \kappa_\xi$. Suppose κ were not a cardinal. Then there would be an ordinal $\alpha < \kappa$ and a surjection

$$f : \alpha \to \kappa.$$

For some $\xi < \delta$, $\alpha < \kappa_\xi$. Define $g : \alpha \to \kappa_\xi$ by

$$g(\nu) = \begin{cases} f(\nu), & \text{if } f(\nu) \in \kappa_\xi, \\ 0, & \text{if } f(\nu) \notin \kappa_\xi. \end{cases}$$

Clearly, g is a surjection, contrary to κ_ξ being a cardinal. Hence κ is a cardinal. □

It follows that the class of all cardinals is in one-one correspondence with the class of all ordinal numbers : there is a proper class of cardinals. Indeed, by the recursion principle the "sequence" (in the class sense) of all infinite cardinal numbers may be defined thus :

$$\omega_0 = \omega$$

$$\omega_{\alpha+1} = \omega_\alpha^+$$

$$\omega_\delta = \lim_{\alpha<\delta} \omega_\alpha, \quad \text{if } \lim(\delta).$$

Now, very shortly we shall define an arithmetic of cardinal numbers, which will not at all resemble the arithmetic we defined for ordinal numbers. Since every cardinal number is, however, an ordinal number, and since we shall use the same symbols +, ., κ^λ as before for the basic arithmetical operations, there arises a possibility of confusion. To try and eliminate this, we adopt the following convention. The symbols

$$\omega_\alpha$$

are to be used whenever we are considering ω_α as an ordinal. If, however, we are using ω_α as a cardinal, we write instead

$$\aleph_\alpha .$$

($\aleph$ is the letter 'aleph' - the first letter of the Hebrew alphabet.) Thus, for example, if we write

$$\omega_\alpha + \omega_\gamma ,$$

it is understood that ordinal addition is meant, whereas

$$\aleph_\alpha + \aleph_\beta$$

will imply cardinal addition.

But bear in mind that the two notations are purely for our convenience. The "equation"

$$\aleph_\alpha = \omega_\alpha$$

is strictly valid. (Indeed, experts in the field often use ω_α at all times, relying on experience to keep out of trouble!)

We are now in a position to give a formal definition to the terms "finite", "countable", "uncountable".

A set is *finite* if its cardinality is less than $\aleph_0$.

A set is *countable* if its cardinality is at most $\aleph_0$.

A set is *uncountable* if its cardinality is at least $\aleph_1$.

Thus, $\aleph_\alpha$ is the α'th uncountable cardinal.

Let us return now to our picture of the ordinal number system. Although we were able to extend this picture quite a way into the transfinite by using our arithmetical notions for ordinals, all of the ordinals considered (even the "giant"

$$\underbrace{\omega^{\omega^{\omega^{\cdot^{\cdot^{\cdot^{\omega^{\cdot^{\cdot^{\cdot}}}}}}}}}}_{\omega \text{ times}} \quad)$$

were countable. (We shall presently be in a position to prove this.) Hence already ω_1 is much bigger. Now we get the more "complete" picture :

$$0, 1, 2, \ldots, \omega, \omega+1, \ldots, \omega.2, \ldots, \omega.3, \ldots, \omega.\omega, \ldots$$
$$\ldots, \omega^3, \ldots, \omega^{\omega}, \ldots, \omega^{\omega^{\omega}}, \ldots \quad \ldots$$
$$\ldots, \omega_1 \ldots, \omega_2, \ldots, \omega_n \ldots, \omega_{\omega}, \ldots$$
$$\ldots, \omega_{\omega_1}, \ldots, \omega_{\omega_2}, \ldots, \omega_{\omega_{\omega}}, \ldots \quad \ldots$$
$$\ldots \quad \ldots \quad .$$

In fact, now we can consider the real "giant"

$$\omega_{\omega_{\omega_{\omega_{\cdot_{\cdot_{\cdot}}}}}} \quad (\omega \text{ times})$$

This cardinal has an interesting property which we study below. First, let us make a rather obvious observation, immediate from the definition.

6.8 Lemma

The function $\aleph : On \to On$ is a normal function. □

It follows that $\omega_\alpha \geq \alpha$ for all α. The ordinal

$$\omega_{\omega_{\omega_{\cdot_{\cdot_{\cdot}}}}} \quad (\omega \text{ times})$$

is the smallest cardinal κ such that $\aleph_\kappa = \kappa$. By 4.7, there is a proper class of such κ. Since the jump from ω_α to $\omega_{\alpha+1}$ is absolutely enormous in the ordinal sense, even though it is only a step of one up in the cardinal sense, the cardinals increase in size way in advance of the ordinals. Nonetheless, there are arbitrarily large cardinals κ which are simultaneously the κ'th ordinal and the κ'th uncountable cardinal : such cardinals being truly "enormous".

Let us end this section with a simple point which rapidly leads into a rather hazardous region.

By our definitions, every set has a unique cardinality. Hence, for each set X there is a unique ordinal α such that $|X| = \aleph_\alpha$. This is known. Calculation of the α involved for particular X is, however, not always easy, and for some sets X the α concerned *cannot* be calculated on the basis of the ZFC axioms alone. But more of that later.

7. Arithmetic of Cardinal Numbers

Let $\langle \kappa_\alpha \mid \alpha < \beta \rangle$ be a sequence of cardinal numbers. The *cardinal sum*

$$\sum_{\alpha<\beta} \kappa_\alpha$$

is defined to be

$$\left| \bigcup_{\alpha<\beta} (\kappa_\alpha \times \{\alpha\}) \right|$$

By manipulation of bijections, it is easily seen that

$$\sum_{\alpha<\beta} \kappa_\alpha = \left| \bigcup_{\alpha<\beta} A_\alpha \right|$$

where $\{A_\alpha \mid \alpha < \beta\}$ is any set of pairwise disjoint sets with $|A_\alpha| = \kappa_\alpha$ for all $\alpha < \beta$.

We write $\kappa_0 + \kappa_1$ in place of $\sum_{\alpha<2} \kappa_\alpha$. Thus

$$\kappa + \lambda = |(\kappa \times \{0\}) \cup (\lambda \times \{1\})|.$$

7.1 Theorem

Let κ, λ, μ be cardinals. Then :

(i) $\kappa + (\lambda + \mu) = (\kappa + \lambda) + \mu$;

(ii) $\kappa + \lambda = \lambda + \kappa$.

And if $\langle \kappa_\alpha \mid \alpha < \beta \rangle$ is any sequence of cardinals and $\langle \lambda_\gamma \mid \gamma < \delta \rangle$ is a rearrangement of this sequence, then

(iii) $\sum_{\alpha<\beta} \kappa_\alpha = \sum_{\gamma<\delta} \lambda_\gamma$.

Proof : Trivial. □

If $\langle A_\alpha | \alpha < \beta\rangle$ is a sequence of sets, the *Cartesian product* of this sequence is the set

$$\mathop{\times}_{\alpha<\beta} A_\alpha = \{ f \mid (f : \beta \to \bigcup_{\alpha<\beta} A_\alpha) \ \& \ (\forall\alpha < \beta)(f(\alpha) \in A_\alpha) \} .$$

If $\langle\kappa_\alpha \mid \alpha < \beta\rangle$ is a sequence of cardinals, the *cardinal product*

$$\prod_{\alpha<\beta} \kappa_\alpha$$

is defined to be

$$\left| \mathop{\times}_{\alpha<\beta} \kappa_\alpha \right| .$$

It is easily seen that

$$\prod_{\alpha<\beta} \kappa_\alpha = \left| \mathop{\times}_{\alpha<\beta} A_\alpha \right|$$

where $\langle A_\alpha \mid \alpha < \beta\rangle$ is any sequence of sets with $|A_\alpha| = \kappa_\alpha$ for all $\alpha < \beta$.

We write $\kappa_0 \cdot \kappa_1$ in place of $\prod_{\alpha<2} \kappa_\alpha$. Since $\mathop{\times}_{\alpha<2} A_\alpha$ is canonically isomorphic to the usual "Cartesian product"

$$A_0 \times A_1 = \{(a_0,a_1) \mid a_0 \in A_0 \wedge a_1 \in A_1\} ,$$

we have

$$\kappa\cdot\lambda = |\kappa \times \lambda| .$$

7.2 <u>Theorem</u>

Let κ, λ, μ be cardinals. Then :

(i) $\kappa.(\lambda.\mu) = (\kappa.\lambda).\mu$;

(ii) $\kappa.\lambda = \lambda.\kappa$.

And if $\langle\kappa_\alpha \mid \alpha < \beta\rangle$ is any sequence of cardinals and $\langle\lambda_\gamma \mid \gamma < \delta\rangle$ is a rearrangement of this sequence, then

(iii) $\prod_{\alpha<\beta} \kappa_\alpha = \prod_{\gamma<\delta} \lambda_\gamma$.

Proof : An easy exercise. □

7.3 <u>Theorem</u>

Let κ, λ, μ be cardinals. Then

$$\kappa.(\lambda + \mu) = \kappa.\lambda + \kappa.\mu .$$

Proof : An easy exercise. □

Thus, cardinal addition and multiplication are commutative and associative, and multiplication distributes over addition. It is also easily seen that $\kappa + \kappa = 2.\kappa$.

If κ, λ are cardinals, the *cardinal power*

$$\kappa^\lambda$$

is defined to be

$$\prod_{\alpha<\lambda} \kappa .$$

It follows at once that

$$\kappa^\lambda = |\{f \mid f : \lambda \to \kappa\}| .$$

We usually write ${}^\lambda\kappa$ for $\{f \mid f : \lambda \to \kappa\}$. With this notation,

$$\kappa^\lambda = |{}^\lambda\kappa| .$$

7.4 Theorem

Let κ, λ, μ be cardinals. Then:

(i) $\kappa^{\lambda}.\kappa^{\mu} = \kappa^{(\lambda+\mu)}$;

(ii) $\kappa^{\lambda}.\mu^{\lambda} = (\kappa.\mu)^{\lambda}$;

(iii) $(\kappa^{\lambda})^{\mu} = \kappa^{(\lambda.\mu)}$.

Proof : As easy exercise. □

Thus, not only do addition and multiplication behave just as in the finite case: so too does exponentiation. Moreover, $\kappa^2 = \kappa.\kappa$, as is easily verified.

Let us also remark that all of these arithmetical notions for arbitrary cardinals reduce to the usual notions in the case where the cardinals are finite. Hence, in the finite case, the cardinal and ordinal arithmetics coincide. But in general these arithmetics are quite distinct. For instance, both cardinal addition and cardinal multiplication are commutative, but neither of the ordinal analogues is commutative.

As we have just seen, the arithmetical operations defined on cardinals above have all the algebraic properties of their finite counterparts. But this does not mean that the arithmetic of infinite cardinals is directly comparable to finite arithmetic. In fact, infinite cardinal arithmetic is essentially trivial, as our next results show.

7.5 Theorem

Let $\kappa \geq \aleph_0$. Then $\kappa.\kappa = \kappa$.

Proof : Suppose not. Let κ be the least infinite cardinal such that $\kappa.\kappa \neq \kappa$. Thus, for all cardinals $\lambda < \kappa$, $\lambda.\lambda < \kappa$.

Let $P = \kappa \times \kappa$. Thus $|P| = \kappa.\kappa > \kappa$. For $\xi < \kappa$, let

$$P_\xi = \{ \langle\alpha,\beta\rangle \in P \mid \alpha + \beta = \xi \}$$

Clearly, $\xi \neq \zeta$ implies $P_\xi \cap P_\zeta = \emptyset$. Moreover, $P = \bigcup_{\xi<\kappa} P_\xi$. For suppose first that $\langle\alpha,\beta\rangle \in P_\xi$. Thus $\alpha + \beta = \xi$, which implies $\alpha, \beta < \kappa$, and hence that $\langle\alpha,\beta\rangle \in P$. Conversely, let $\alpha, \beta < \kappa$. Thus $|\alpha|, |\beta| < \kappa$. Hence by choice of κ, $\lambda.\lambda < \kappa$, where $\lambda = \max(|\alpha|, |\beta|)$. But $|\alpha+\beta| = |\alpha| + |\beta| \leq \lambda + \lambda = 2.\lambda \leq \lambda.\lambda < \kappa$. Hence $\alpha + \beta < \kappa$. Thus, setting $\xi = \alpha + \beta$, we get $\langle\alpha,\beta\rangle \in P_\xi$. Thus P_ξ, $\xi < \kappa$, constitutes a partition of P.

Define a well-ordering $<_\xi$ of P_ξ now by

$$\langle\alpha,\beta\rangle <_\xi \langle\alpha',\beta'\rangle \text{ iff } (\alpha<\alpha') \vee (\alpha=\alpha' \wedge \beta<\beta').$$

Define a well-ordering $<_*$ of P by

$$\langle\alpha,\beta\rangle <_* \langle\alpha',\beta'\rangle \text{ iff } [\langle\alpha,\beta\rangle \in P_\xi \wedge \langle\alpha',\beta'\rangle \in P_{\xi'} \wedge \xi<\xi']$$

$$\vee [\langle\alpha,\beta\rangle, \langle\alpha',\beta'\rangle \in P_\xi \wedge \langle\alpha,\beta\rangle <_\xi \langle\alpha',\beta'\rangle].$$

Let $\theta = \mathrm{Ord}(\langle P, <_*\rangle)$. Since $|P| > \kappa$, $\theta > \kappa$. It follows that there is a point $\langle\alpha_0,\beta_0\rangle$ in P such that $\mathrm{Ord}(\langle Q, <_*\rangle) = \kappa$, where

$$Q = \{ \langle\alpha,\beta\rangle \in P \mid \langle\alpha,\beta\rangle <_* \langle\alpha_0,\beta_0\rangle \}.$$

Pick $\xi_0 < \kappa$ with $\langle\alpha_0, \beta_0\rangle \in P_{\xi_0}$. Thus $\alpha_0 + \beta_0 = \xi_0$. Then

$$\langle\alpha,\beta\rangle \in Q \rightarrow \langle\alpha, \beta\rangle <_* \langle\alpha_0, \beta_0\rangle$$

$$\rightarrow \alpha,\beta \leq \xi_0 .$$

Hence $$Q \subseteq (\xi_0 + 1) \times (\xi_0 + 1) .$$

But $\xi_0 + 1 < \kappa$, so $|\xi_0 + 1| < \kappa$ so we have

$$|Q| \leq |\xi_0 + 1|.|\xi_0 + 1| < \kappa ,$$

contrary to $\mathrm{Ord}(\langle Q, <_*\rangle) = \kappa$. The proof is complete. □

7.6 Corollary

Let κ, λ be cardinals, $\kappa \leq \lambda$, $\lambda \geq \aleph_o$. Then $\kappa.\lambda = \kappa+\lambda = \lambda$.

Proof : $\lambda \leq \kappa + \lambda \leq \lambda + \lambda = 2.\lambda \leq \lambda.\lambda = \lambda$

and $\lambda \leq \kappa.\lambda \leq \lambda.\lambda = \lambda$. □

7.7 Corollary

Let $\kappa \geq \aleph_o$. Then $\kappa^+ = |\{ \alpha \mid \kappa \leq \alpha < \kappa^+ \}|$. (i.e. the set of all ordinals of cardinality κ has cardinality κ^+.)

Proof : $\kappa^+ = |\{ \alpha \mid \alpha < \kappa^+ \}|$

$= |\{ \alpha \mid \alpha < \kappa \} \cup \{ \alpha \mid \kappa \leq \alpha < \kappa^+ \}|$

$= |\{ \alpha \mid \alpha < \kappa \}| + |\{ \alpha \mid \kappa \leq \alpha < \kappa^+ \}|$

$= \kappa + |\{ \alpha \mid \kappa \leq \alpha < \kappa^+ \}|$.

By 7.6, we must have $\kappa^+ = |\{ \alpha \mid \kappa \leq \alpha < \kappa^+ \}|$. □

7.8 Corollary

The union of at most κ sets of cardinality at most κ has cardinality at most κ, for any infinite cardinal κ. In particular, the union of countably many countable sets is countable.

Proof : If $|A_\alpha| \leq \kappa$ for each $\alpha < \lambda$, where $\lambda \leq \kappa$, then

$$\left| \bigcup_{\alpha<\lambda} A_\alpha \right| \leq \kappa.\lambda \leq \kappa.\kappa = \kappa.$$ □

7.9 Corollary

For any α, β, $\aleph_\alpha + \aleph_\beta = \aleph_\alpha \cdot \aleph_\beta = \aleph_{\max(\alpha,\beta)}$.

Proof : Immediate. □

7.10 Corollary

For any α, $\aleph_\alpha = \sum_{\beta \le \alpha} \aleph_\beta$.

Proof : Let $Z_\beta = \{ \xi \mid \omega_\beta \le \xi < \omega_{\beta+1} \}$ for each $\beta < \alpha$.

Then $\omega_\alpha = \omega \cup (\bigcup_{\beta<\alpha} Z_\beta)$.

Using 7.7, $\aleph_\alpha = |\omega_\alpha| = \aleph_0 + \sum_{\beta<\alpha} \aleph_{\beta+1} \ge \aleph_0 + \sum_{\beta<\alpha} \aleph_\beta$.

So, by 7.6, $\aleph_\alpha = (\aleph_0 + \sum_{\beta<\alpha} \aleph_\beta) + \aleph_\alpha$

$= \sum_{\beta \le \alpha} \aleph_\beta$. □

7.11 Corollary

If $\lim(\alpha)$, then $\aleph_\alpha = \sum_{\beta<\alpha} \aleph_\beta$.

Proof : Arguing as in 7.10, we get

$$\aleph_\alpha = \aleph_0 + \sum_{\beta<\alpha} \aleph_{\beta+1} .$$

But since $\lim(\alpha)$, $\sum_{\beta<\alpha} \aleph_{\beta+1} = \sum_{\beta<\alpha} \aleph_\beta$.

Hence, using 7.6, $\aleph_\alpha = \sum_{\beta<\alpha} \aleph_\beta$. □

Cardinal exponentiation turns out to be much more difficult to handle than addition and multiplication, so we shall postpone our discussion of this topic until later (section 9), and end our present investigation of cardinal arithmetic where it stands.

8. Cofinality. Singular and Regular Cardinals

A cardinal of the form κ^+ is called a *successor cardinal*. For instance, 1, 2, 3, ... are all successor cardinals. So too are $\aleph_1$, $\aleph_2$, $\aleph_3$. Indeed, an infinite cardinal will be a successor cardinal iff it is of the form $\aleph_{\alpha+1}$ for some α; or, to put it another way, a cardinal $\aleph_\gamma$ is a successor *cardinal* iff the index γ is a successor *ordinal*.

A cardinal which is not a successor cardinal is called a *limit cardinal*. Examples of limit cardinals are 0, $\aleph_0$, $\aleph_\omega$, $\aleph_{\omega+\omega}$, $\aleph_{\omega.\omega}$, $\aleph_{\omega_1}$. Indeed, $\aleph_\gamma$ will be a limit *cardinal* just in case γ is a limit *ordinal*. The properties of limit cardinals, including many of their arithmetical properties, is closely bound up with a property known as *cofinality*, which we examine now.

Let λ be a limit ordinal. A set $A \subseteq \lambda$ is said to be *bounded* in λ iff there is a $\gamma < \lambda$ such that $A \subseteq \gamma$; otherwise A is *unbounded* in λ. Thus $A \subseteq \lambda$ is unbounded in λ iff

$$(\forall\alpha \in \lambda)(\exists\beta \in A)(\beta > \alpha).$$

Now let θ be a limit ordinal, $\langle\gamma_\nu \mid \nu < \theta\rangle$ an increasing sequence of ordinals in λ. We say $\langle\gamma_\nu \mid \nu < \theta\rangle$ is *cofinal* in λ iff $\{\gamma_\nu \mid \nu < \theta\}$ is an unbounded subset of λ.

8.1 Lemma

$\langle\gamma_\nu \mid \nu < \theta\rangle$ is cofinal in λ iff $\bigcup_{\nu<\theta} \gamma_\nu = \lambda$.

Proof : Trivial. □

The *cofinality* of λ, $\mathrm{cf}(\lambda)$, is the least limit ordinal θ such that there exists an increasing θ-sequence which is cofinal in λ. We shall shortly be able to discuss several examples of this concept. But first some simple lemmas and some further definitions.

8.2 Lemma

$cf(\lambda)$ is a cardinal.

Proof : An easy exercise. □

8.3 Lemma

If $cf(\lambda) = \theta$, there is an increasing θ-sequence, cofinal in λ, which is continuous at every limit ordinal in θ.

Proof : An easy exercise. □

An infinite cardinal κ is *regular* iff $cf(\kappa) = \kappa$; otherwise κ is *singular*. Thus κ is singular iff $cf(\kappa) < \kappa$.

For instance, $\aleph_0$ is clearly regular. And since $cf(\aleph_\omega) = \omega$ (consider the sequence $\langle\omega_n \mid n < \omega\rangle$), $\aleph_\omega$ is singular.

8.4 Lemma

For any limit ordinal λ, $cf(\lambda)$ is a regular cardinal.

Proof : An easy exercise. □

The following theorem relates the notion of cofinality to cardinal arithmetic.

8.5 Theorem

Let κ be an infinite cardinal. Let θ be the least ordinal such that there is a sequence $\langle\kappa_\nu \mid \nu < \theta\rangle$ of cardinals $\kappa_\nu < \kappa$ with

$$\kappa = \sum_{\nu<\theta} \kappa_\nu .$$

Then $\theta = cf(\kappa)$.

Proof : Let $\lambda = cf(\kappa)$. Notice that by 7.9, $\theta \geq \omega$, and by 7.1(iii) θ is a cardinal. Suppose first that $\theta < \lambda$. Thus, for some $\gamma < \kappa$, $\{\kappa_\nu \mid \nu < \theta\} \subseteq \gamma$. Then, $\kappa_\nu \leq |\gamma|$ for all $\nu < \theta$, so

$$\kappa = \sum_{\nu<\theta} \kappa_\nu \leq \sum_{\nu<\theta} |\gamma| = \theta.|\gamma| = \max(\theta,|\gamma|) < \kappa ,$$

which is absurd.

Now suppose that $\lambda < \theta$. By 8.3, let $\langle\gamma_\nu \mid \nu < \lambda\rangle$ be a normal sequence cofinal in κ. We may assume that $\gamma_0 = 0$. Then, by 8.1, $\kappa = \bigcup_{\nu<\lambda} \gamma_\nu = \bigcup_{\nu<\lambda} (\gamma_{\nu+1} - \gamma_\nu)$. Setting $\mu_\nu = |\gamma_{\nu+1} - \gamma_\nu|$, we get $\kappa = |\kappa| = \sum_{\nu<\lambda} \mu_\nu$, contrary to $\lambda < \theta$. □

Using 8.5, we can now show that there are many regular cardinals.

8.6 Theorem

Every infinite successor cardinal is regular.

Proof : Let κ be any infinite cardinal. We show that κ^+ is regular. Let $\lambda = cf(\kappa^+)$. By 8.5 we can find cardinals $\kappa_\nu < \kappa^+$, $\nu < \lambda$, such that

$$\kappa^+ = \sum_{\nu<\lambda} \kappa_\nu .$$

For each ν, $\kappa_\nu \leq \kappa$, so

$$\kappa^+ \leq \sum_{\nu<\lambda} \kappa = \lambda.\kappa .$$

Hence $\lambda = \kappa^+$. □

8.7 Corollary

Every singular cardinal is a limit cardinal. □

In section 10 we discuss the converse to 8.7. Meanwhile, let us consider a few examples. By 8.6, $\aleph_1$, $\aleph_2$, $\aleph_3$, ... are all regular. And $\aleph_\omega$ is singular, of cofinality ω. $\aleph_{\omega+1}$, $\aleph_{\omega+2}$, ... are all regular. And $\aleph_{\omega+\omega}$ is singular of cofinality ω. $\aleph_{\omega_1}$ is singular of cofinality ω_1; $\aleph_{\omega_2}$ is singular of cofinality ω_2, etc. And as a general lemma, we have :

8.8 Lemma

$\lim(\alpha) \rightarrow \mathrm{cf}(\aleph_\alpha) = \mathrm{cf}(\alpha)$.

Proof : Trivial. □

In section 9 we shall meet cases where the cardinal arithmetic is affected by the cofinality of the cardinals concerned. As a first example of cofinality properties, however, let us consider 4.7, the fixed point theorem for normal functions. Previously we were only able to obtain this result for class functions $f : \mathrm{On} \rightarrow \mathrm{On}$. We may now generalise this result.

8.9 Theorem (Fixed Point Theorem for Normal Functions)

Let $\lim(\lambda)$, $\mathrm{cf}(\lambda) > \omega$. If $f : \lambda \rightarrow \lambda$ is a normal function, then for every $\alpha \in \lambda$ there is a fixed point γ of f such that $\gamma \geq \alpha$.

Proof : Let $\alpha \in \lambda$ be given. If $f(\alpha) = \alpha$ there is nothing further to prove. So assume otherwise. By 4.6, $f(\alpha) > \alpha$. By recursion, define a function $g : \omega \rightarrow \lambda$ so that $g(0) = 0$ and $g(n+1) = f(g(n))$. By induction, $g(n) < g(n+1)$ and g maps into λ. Let $\gamma = \lim_{n<\omega} g(n)$. Since $\mathrm{cf}(\lambda) > \omega$, g cannot be cofinal in λ, so $\gamma < \lambda$. But clearly, $f(\gamma) = \gamma$. □

9. Cardinal Exponentiation

We consider now the function κ^λ. First a useful characterisation of 2^κ.

9.1 Lemma

For any cardinal κ, $2^{\kappa} = |\mathcal{P}(\kappa)|$.

Proof : By definition, $2^{\kappa} = |{}^{\kappa}2| = |\{ f \mid f : \kappa \to 2 \}|$. But, of course, there is a well-known one-one correspondence between the sets $\{ f \mid f : \kappa \to 2 \}$ and $\mathcal{P}(\kappa)$: we associate with each set $X \subseteq \kappa$ its characteristic function $\chi_X : \kappa \to 2$ defined by

$$\chi_X(\xi) = 1 \leftrightarrow \xi \in X. \qquad \square$$

9.2 Corollary

For any cardinal κ, $2^{\kappa} > \kappa$.

Proof : The proof of 6.6 shows that $|\mathcal{P}(\kappa)| > \kappa$. $\square$

9.3 Theorem

Let κ, λ be cardinals, λ infinite, $\kappa \leq \lambda$. Then $\kappa^{\lambda} = 2^{\lambda}$.

Proof : Clearly, $2^{\lambda} \leq \kappa^{\lambda}$. We show that $\kappa^{\lambda} \leq 2^{\lambda}$. Since λ is infinite and $\kappa \leq \lambda$, $\kappa.\lambda = \lambda$. Let $j : \lambda \times \kappa \leftrightarrow \lambda$. For each function $h : \lambda \to \kappa$, we have (formally) $h \subseteq \lambda \times \kappa$. Set $G(h) = j[h]$. Thus $G(h) \subseteq \lambda$. Clearly, $G : {}^{\lambda}\kappa \to \mathcal{P}(\lambda)$ is an injection. Hence, using 9.1,

$$\kappa^{\lambda} = |{}^{\lambda}\kappa| \leq |\mathcal{P}(\lambda)| = 2^{\lambda}. \qquad \square$$

Hence, if λ is infinite, the behaviour of κ^{λ} as κ varies up to λ is known. For $\kappa > \lambda$, the picture becomes more complex. We have, for example:

9.4 Theorem

Let κ be an infinite cardinal. Then $(\kappa^{+})^{\kappa} = 2^{\kappa}$.

Proof : Clearly, $\kappa^+.2^\kappa \leq (\kappa^+)^\kappa$. By 8.6, κ^+ is regular, so

$$^\kappa(\kappa^+) \subseteq \bigcup_{\alpha<\kappa^+} {}^\kappa\alpha$$

which gives

$$\begin{aligned}(\kappa^+)^\kappa &\leq \sum_{\alpha<\kappa^+} |{}^\kappa\alpha| \\ &= \sum_{\alpha<\kappa^+} |\alpha|^\kappa \\ &= \kappa^+.\kappa^\kappa \\ &= \kappa^+.2^\kappa = 2^\kappa . \quad \square\end{aligned}$$

But in general there is little more that can be said. Indeed, although we know that there is an α such that $2^{\aleph_0} = \aleph_\alpha$, the ZFC axioms do not provide us with enough information to decide *which* α solves this equation. The "obvious" guess is, perhaps, that $2^{\aleph_0} = \aleph_1$. This was already proposed by Cantor at the beginning of this century. By considering the representations of the real numbers in the unit interval (0, 1) as non-terminating decimal expansions, one sees easily that the cardinality of this interval is $10^{\aleph_0} = 2^{\aleph_0}$. Since (0, 1) is known to be homeomorphic to the whole real line, $\mathbb{R}$, it follows that $|\mathbb{R}| = 2^{\aleph_0}$. Hence the question as to which solves $2^{\aleph_0} = \aleph_\alpha$ can be expressed thus : How many real numbers are there? Expressed in this manner, the question became known as the *continuum problem*. The hypothesis $2^{\aleph_0} = \aleph_1$ of Cantor (i.e. $|\mathbb{R}| = \aleph_1$) became known as the *continuum hypothesis* (CH for short). Despite the relative ease with which CH can be stated, and despite its extreme simplicity, efforts over the years to resolve the continuum problem met with no success. In the 1930's, Kurt Gödel used techniques of logic to show that CH could certainly not be disproved (on the basis of the ZFC axioms), but a moment's thought will indicate that this does not in itself *prove* the CH. It could be that the continuum problem is not decidable on the basis of our axioms. And, as Paul Cohen showed in 1963, this is indeed the case. The ZFC axioms do not resolve the continuum problem, and indeed do not resolve very many of the questions one can raise about cardinal exponentiation. We shall try to provide some explanation as to *why* CH is not decidable in the ZFC system in Chapter 5. And in Chapter 6 we give some indication of the methods by which it can be *proved* that a particular statement, such as CH, is not decidable in the ZFC system. In the meantime, in order not to leave any loose ends, let us present the one and only positive result about the size of $2^{\aleph_0}$ which we have (and indeed can ever have). We can prove that, for instance, $2^{\aleph_0}$ cannot equal $\aleph_\omega$; or $\aleph_{\omega+\omega}$; or indeed any cardinal of cofinality ω. In order to prove this, we first establish a very general result on cardinal arithmetic.

9.5 Theorem

Let κ_α, λ_α be cardinals, $\alpha < \beta$, with $\kappa_\alpha < \lambda_\alpha$ for all $\alpha < \beta$. Then

$$\sum_{\alpha<\beta} \kappa_\alpha < \prod_{\alpha<\beta} \lambda_\alpha$$

Proof : Define $f : \bigcup_{\alpha<\beta} (\kappa_\alpha \times \{\alpha\}) \to \bigtimes_{\alpha<\beta} \lambda_\alpha$ by letting $f(\xi, \gamma)$ be that element of $\bigtimes_{\alpha<\beta} \lambda_\alpha$ which takes the value $\xi \in \lambda_\gamma$ in the γ-th place, and the value 0 elsewhere. That is

$$f(\xi, \gamma)(\nu) = \begin{cases} \xi, & \text{if } \nu = \gamma , \\ 0, & \text{otherwise} . \end{cases}$$

Clearly, f is an injection. Hence

$$\sum_{\alpha<\beta} \kappa_\alpha = \left| \bigcup_{\alpha<\beta} (\kappa_\alpha \times \{\alpha\}) \right| \le \left| \bigtimes_{\alpha<\beta} \lambda_\alpha \right| = \prod_{\alpha<\beta} \lambda_\alpha .$$

Suppose that $\sum_{\alpha<\beta} \kappa_\alpha = \prod_{\alpha<\beta} \lambda_\alpha$. Let

$$f = \bigcup_{\alpha<\beta} (\kappa_\alpha \times \{\alpha\}) \xrightarrow{\text{onto}} \bigtimes_{\alpha<\beta} \lambda_\alpha .$$

For $\alpha < \beta$, let f_α be the projection of f onto λ_α; that is

$$f_\alpha(\xi, \gamma) = f(\xi, \gamma)(\alpha)$$

Then $$f_\alpha \restriction (\kappa_\alpha \times \{\alpha\}) : \kappa_\alpha \times \{\alpha\} \to \lambda_\alpha .$$

Since $\kappa_\alpha < \lambda_\alpha$ and $|\kappa_\alpha \times \{\alpha\}| = \kappa_\alpha$, $f_\alpha \restriction (\kappa_\alpha \times \{\alpha\})$ cannot be surjective. Hence we can pick $\delta_\alpha \in \lambda_\alpha - f_\alpha[\kappa_\alpha \times \{\alpha\}]$. Let $\sigma = \langle\delta_\alpha \mid \alpha < \beta\rangle$. Then $\sigma \in \bigtimes_{\alpha<\beta} \lambda_\alpha$, so for some ξ, α, $\sigma = f(\xi, \alpha)$. Thus, in particular, $\delta_\alpha = f(\xi, \alpha)(\alpha) = f_\alpha(\xi, \alpha) \in f_\alpha[\kappa_\alpha \times \{\alpha\}]$, which is absurd. The theorem is proved. □

9.6 <u>Corollary</u>

For any infinite cardinal κ, $\kappa^{cf(\kappa)} > \kappa$.

Proof : Let $\lambda = cf(\kappa)$. By 8.5 we can find cardinals $\kappa_\alpha < \kappa$, $\alpha < \lambda$, such that $\kappa = \sum_{\alpha<\lambda} \kappa_\alpha$. Since $\kappa_\alpha < \kappa$ for all $\alpha < \lambda$, 9.5 gives $\kappa = \sum_{\alpha<\lambda} \kappa_\alpha < \prod_{\alpha<\lambda} \kappa = \kappa^\lambda$, as required. □

And as a consequence of 9.6, we have our result on $2^{\aleph_0}$:

9.7 <u>Theorem</u>

For any infinite cardinal κ, $cf(2^\kappa) > \kappa$. Hence, in particular, $cf(2^{\aleph_0}) > \omega$ (giving $2^{\aleph_0} \neq \aleph_\omega$, etc.)

Proof : Suppose $cf(2^\kappa) \leq \kappa$. Then, setting $\lambda = 2^\kappa$, we get

$$\lambda^{cf(\lambda)} = \lambda^{cf(2^\kappa)} \leq \lambda^\kappa = (2^\kappa)^\kappa = 2^{\kappa\cdot\kappa} = 2^\kappa = \lambda ,$$

contrary to 9.6 (for λ). □

10. <u>Inaccessible Cardinals</u>

In 8.7 we proved that every singular cardinal is a limit cardinal. What about the converse? Well, $\aleph_0$ is a regular limit cardinal. But are there any others?

If you attempt to find any you will (for certain) fail. Any uncountable limit cardinal which one can "construct" in the ZFC system is singular. Or, to put it another way, in ZFC it is not possible to prove that there is an uncountable regular limit cardinal. On the other hand, it is extremely unlikely that one could prove (in ZFC) that *no such* cardinal can exist : indeed, though the existence of an uncountable regular limit cardinal cannot be proved in ZFC, it is arguable that the existence of such cardinals is implicit in the motivation leading to the ZFC axioms. Accordingly we give such cardinals a name - *weakly inaccessible cardinals* - and study them. There are at least three reasons for doing this.† Firstly, suppose we are trying to prove some mathematical result by an induction on cardinals. (There are many instances of such proofs, in most areas of mathematics.) The induction step at a singular limit cardinal will often make use of properties peculiar to singular cardinals. For the induction step at regular limit cardinals - weakly inaccessible cardinals - a separate argument is then used, making use of properties of weakly inaccessible cardinals. For which purpose a study of such cardinals is necessary, leaving aside questions as to whether weakly inaccessible cardinals "exist". Our second reason for looking at weakly inaccessible cardinals will appeal perhaps only to the set theorist. The existence of such cardinals cannot be proved, but it is (arguably) inherent in the basic ideas of set theory (see later); and hence the notion of an inaccessible cardinal is worthy of study as an aspect of pure set theory. Of course, to do so one needs to adjoin to the ZFC system an axiom which asserts the existence of a weakly inaccessible cardinal. This is directly analogous to the inclusion of the Axiom of Infinity in the ZFC system. Without this axiom, one cannot prove (in ZFC) the existence of an infinite set. Because we want infinite sets, we introduce an axiom which gives us one. Weakly inaccessible cardinals are just uncountable analogues of $\aleph_0$ (recall that $\aleph_0$ is a regular limit cardinal). Which brings us to the argument that weakly inaccessible cardinals are inherent in our intuition on set theory. The cardinal

---

†A fourth, and perhaps more unexpected reason will be given in IV.5. The existence of weakly inaccessible cardinals actually follows from a rather natural measure theoretic assumption.

$\aleph_0$, being both regular and a limit cardinal, is very much larger than any of its predecessors. Neither the replacement axiom nor the cardinal successor function gets up to $\aleph_0$ from below. But our set theoretic inverse should surely possess a uniform character! The cardinal $\aleph_0$ should not be so unusual. There ought to be a proper class of such cardinals. In fact, we can take this a step further. Let us call an uncountable cardinal κ *strongly inaccessible* (or just *inaccessible*) if it is regular and $(\forall\lambda<\kappa)(2^{\lambda}<\kappa)$. Clearly, every strongly inaccessible cardinal is weakly inaccessible. (We discuss the converse later.) Also, apart from not being uncountable, $\aleph_0$ is strongly inaccessible. An inaccessible cardinal is one which cannot be constructed using any of the ZFC axioms : in particular, its definition precludes construction using the axioms of replacement and power set, the two powerful construction axioms which provided all of our cardinal existence results hitherto. Since $\aleph_0$ cannot be constructed using the ZFC axioms (without the Axiom of Infinity), uniformity demands (?) the existence of a proper class of such cardinals. In fact, one rarely makes any fuss about this, because the adjunction of "many" inaccessible cardinal axioms to the ZFC system does not seem to increase the richness of the set theory very much. At most one usually just studies one or two inaccessible cardinals. And so to our third reason. As we mentioned, inaccessible cardinals and weakly inaccessible cardinals resemble $\aleph_0$ to some extent. And we all feel that there is some fundamental difference between finite and infinite which is not shared by any other division, such as between countable and uncountable.† Finiteness is a very special property. By studying inaccessibility properties, one can hope to gain some insight into how "finiteness" behaves, without getting at all involved in the finite itself.

But now it is high time that we got down to the business of looking at inaccessible cardinals. We commence by re-stating the definitions.

A cardinal κ is *weakly inaccessible* iff it is an uncountable, regular limit cardinal.

---

†A little reflection should indicate that this is not at variance with our above discussion about the uniformity of the universe.

A cardinal κ is *(strongly) inaccessible* iff it is an uncountable regular cardinal and $(\forall\lambda < \kappa)(2^\lambda < \kappa)$.

Clearly, every inaccessible cardinal is weakly inaccessible. The converse is not in general true. The *Generalised Continuum Hypothesis* is the assertion

$$(\forall\kappa \geq \aleph_0)[2^\kappa = \kappa^+]$$

(i.e. $\forall\alpha[2^{\aleph_\alpha} = \aleph_{\alpha+1}]$), denoted by GCH. It is known to be consistent with the ZFC axioms. Clearly, if we assume GCH (as an additional axiom of set theory), then the notions of weak inaccessibility and inaccessibility coincide. But it is also consistent with the ZFC axioms that these two notions are quite different. Indeed, it is not possible to prove in ZFC that $2^{\aleph_0}$ is not weakly inaccessible, though it is trivially not inaccessible.

10.1 Lemma

If $\aleph_\kappa$ is a weakly inaccessible cardinal, then $\aleph_\kappa = \kappa$.

Proof : Suppose $\kappa \neq \aleph_\kappa$. Thus $\kappa < \aleph_\kappa$. Since κ is a limit ordinal, $cf(\aleph_\kappa) = cf(\kappa) < \aleph_\kappa$, contrary to $\aleph_\kappa$ being regular. □

Our next result gives an application of inaccessibility to cardinal arithmetic.

10.2 Theorem

If κ is an inaccessible cardinal, then $\sum_{\lambda<\kappa} \kappa^\lambda = \kappa$.

Proof : Assume κ is inaccessible. Since κ is regular, if $\lambda < \kappa$, then

$$^\lambda\kappa = \bigcup_{\alpha<\kappa} {}^\lambda\alpha .$$

Hence, for $\lambda < \kappa$,

$$\kappa^\lambda \leq \sum_{\alpha<\kappa} |\alpha|^\lambda .$$

But if λ, $\alpha < \kappa$, then since κ is inaccessible, $|\alpha|^\lambda < \kappa$. Hence, for $\lambda < \kappa$,

$$\kappa^{\lambda} \leq \sum_{\alpha<\kappa} \kappa = \kappa.\kappa = \kappa .$$

Thus, $$\sum_{\lambda<\kappa} \kappa^{\lambda} \leq \sum_{\lambda<\kappa} \kappa = \kappa.\kappa = \kappa .$$

The theorem is proved. □

Assuming GCH (as an additional axiom of set theory) we can extend 10.2 thus:

10.3 <u>Theorem</u>

Assume GCH. Let κ be a limit cardinal. Then κ is inaccessible iff

$$\sum_{\lambda<\kappa} \kappa^{\lambda} = \kappa .$$

Proof : 10.2 gives one half of the result. So let us suppose κ is not inaccessible. By GCH, κ is not weakly inaccessible. So, being a limit cardinal, κ must be singular. Thus $cf(\kappa) < \kappa$. But $\kappa^{cf(\kappa)} > \kappa$, by 9.6. Hence $\sum_{\lambda<\kappa} \kappa^{\lambda} > \kappa$. □

The most significant fact concerning inaccessible cardinals is the following, which we do not prove here.† If κ is inaccessible, then the set V_{κ} is a "fixed point" or "closure point" in the cumulative hierarchy with respect to the ZFC axioms. That is, if we use the ZFC axioms in any manner to define new sets from sets in V_{κ}, then these new sets will lie in V_{κ} : the axioms of ZFC will not lead out of V_{κ}! Consequently, V_{κ} is a "model" of the ZFC axioms.

---

†This brief discussion is not very precise, and will not stand up to much detailed consideration. To make it precise would, however, lead us far from our goal, so we content ourselves with an evocative, if not totally rigorous remark.

Problems

1. (Ordinal Arithmetic) The operations of ordinal addition and multiplication may be defined by recursion.

(A) For each α, define a function $a_\alpha : On \to On$ by the recursion

$$a_\alpha(0) = \alpha ;$$

$$a_\alpha(\beta+1) = a_\alpha(\beta) + 1 ;$$

$$a_\alpha(\gamma) = \bigcup_{\beta<\gamma} a_\alpha(\beta), \text{ if } \lim(\gamma).$$

Then $a_\alpha(\beta) = \alpha + \beta$.

(B) For each α, define a function $m_\alpha : On \to On$ by the recursion

$$m_\alpha(0) = 0 ;$$

$$m_\alpha(\beta+1) = m_\alpha(\beta) + \alpha ;$$

$$m_\alpha(\gamma) = \bigcup_{\beta<\gamma} m_\alpha(\beta), \text{ if } \lim(\gamma).$$

Then $m_\alpha(\beta) = \alpha.\beta$.

(C) By proving an ordinal recursion principle which allows for parameters, it is possible to modify the above definitions to obtain recursive definitions of + and . as functions from On × On to On.

2. (Cardinal Arithmetic)

(A) Let κ_α, $\alpha < \beta$, be infinite cardinals with $\kappa = \sum_{\alpha<\beta} \kappa_\alpha$. Then for any cardinal λ, $\lambda^\kappa = \prod_{\alpha<\beta} \lambda^{\kappa_\alpha}$.

(B) If κ is regular and $2^\mu \leq \kappa$ for all $\mu < \kappa$, then $\sum_{\mu<\kappa} \kappa^\mu = \kappa$.

(C) Assume GCH. Then $\kappa^\lambda = \kappa$ for all $\lambda < cf(\kappa)$. If κ is regular, then $\sum_{\mu<\kappa} \kappa^\mu = \kappa$.

(D) For any infinite κ, $\kappa^\kappa \geq \sum_{\lambda<\kappa} \kappa^\lambda \geq \sum_{\lambda<\kappa} 2^\lambda \geq \kappa$.

(E) If $\kappa = \lambda^+$, then $\sum_{\mu<\kappa} \kappa^\mu = \sum_{\mu<\kappa} 2^\mu = 2^\lambda$.

(F) GCH is equivalent to the condition $\sum_{\mu<\kappa} 2^\mu = \kappa$ for all infinite κ.

(G) GCH is equivalent to the condition $\kappa^{cf(\kappa)} = \kappa^+$ for all infinite κ.

(H) $\sum_{\lambda<\kappa} \kappa^\lambda = \kappa$ iff κ is regular and $\sum_{\lambda<\kappa} 2^\lambda = \kappa$.

3. (The Order Topology on ω_1) The order topology has been defined in Problems I.3. Here we consider the particular spaces given by the order topology on ω_1 and $\omega_1 + 1$.

(A) ω_1 is first countable but not second countable. $\omega_1 + 1$ is not even first countable.

(B) Let $A \subseteq \omega_1$. A is *closed* iff, whenever γ is a limit ordinal in ω_1 such that $A \cap \gamma$ is unbounded in γ, then $\gamma \in A$. Similarly with $\omega_1 + 1$ in place of ω_1.

(C) Let $\alpha \in \omega_1$. $\{\alpha\}$ is open (i.e. α is isolated) iff α is a successor ordinal.

(D) Both ω_1 and $\omega_1 + 1$ are Hausdorff spaces.

(E) ω_1 is locally compact but not compact. $\omega_1 + 1$ is not locally compact.

(F) Both ω_1 and $\omega_1 + 1$ are Lindelöf.

(G) Both ω_1 and $\omega_1 + 1$ are normal.

(H) $\omega_1 \times (\omega_1 + 1)$ is not normal.

Chapter IV. Some Topics in Pure Set Theory†

1. The Borel Hierarchy

Recall the definition of a *Borel set*. If $\mathcal{H}$ is any collection of subsets of a set X, there is a unique smallest σ-field of subsets of X containing $\mathcal{H}$: the σ-field *generated* by $\mathcal{H}$. (The notion of a field of sets was defined in Problem I.1. A field, $\mathcal{F}$, of subsets of X is called a σ-field iff whenever $A_n \in \mathcal{F}$, $n = 1,2,\ldots,$ then $\bigcup_{n=1}^{\infty} A_n \in \mathcal{F}$ and $\bigcap_{n=1}^{\infty} A_n \in \mathcal{F}$.) In $\mathbb{R}$, the σ-field generated by the collection of open sets is the σ-field of all *Borel sets*: a subset of $\mathbb{R}$ is a *Borel set* iff it is a member of this σ-field. We show that the Borel sets lie in a ramified hierarchy, and that there are precisely $2^{\aleph_0}$ many Borel sets.

Let B_0 denote the set of all open intervals (a, b), $a, b \in \mathbb{R}$. And by recursion, let $B_{\alpha+1}$ consist of all sets of reals which are either a countable union of members of B_α, or a countable intersection of members of B_α, or the difference of two members of B_α, with $B_\lambda = \bigcup_{\alpha<\lambda} B_\alpha$ if $\lim(\lambda)$. Clearly, $\nu < \tau \rightarrow B_\nu \subseteq B_\tau$. Set

$$B = \bigcup_{\alpha<\omega_1} B_\alpha .$$

---

†This chapter contains some fairly complex proofs, which the casual reader can ignore. Chapter V and VI do not depend upon this chapter. For the reader who is interested in seeing what arguments in set theory look like, let us mention that, unless stated otherwise, the sections in this chapter are all independent of each other.

1.1 Theorem

B is the set of all Borel sets.

Proof : Let $\mathfrak{F}$ be the σ-field of all Borel sets. By induction on $\alpha < \omega_1$, we prove $B_\alpha \subseteq \mathfrak{F}$. For $\alpha = 0$ this is clear. And the induction step at limit ordinals is trivial. Now suppose $B_\alpha \subseteq \mathfrak{F}$. If $X \in B_{\alpha+1}$, then X is either a countable union of members of $\mathfrak{F}$, or a countable intersection of members of $\mathfrak{F}$, or the difference of two members of $\mathfrak{F}$. Hence as $\mathfrak{F}$ is a σ-field (and thus closed under such operations), $X \in \mathfrak{F}$. Thus $B \subseteq \mathfrak{F}$. To prove the reverse inclusion, $\mathfrak{F} \subseteq B$, it suffices to prove that B is a σ-field of subsets of $\mathbb{R}$ which contains all open sets. (Because $\mathfrak{F}$ is the smallest such σ-field.) Well, since every open set can be expressed as a countable union of open intervals, already B_1 contains all open sets. We show that B is a σ-field.

Suppose $X, Y \in B$. Then for some $\alpha, \beta < \omega_1$, $X \in B_\alpha$, $Y \in B_\beta$. Let $\gamma = \max(\alpha, \beta)$. Then $X, Y \in B_\gamma$, so $X - Y \in B_{\gamma+1} \subseteq B$.

Now suppose $X_n \in B$, $n = 1, 2, \ldots$. For each n, pick $\alpha_n < \omega_1$ so that $X_n \in B_{\alpha_n}$. Since $\mathrm{cf}(\omega_1) > \omega$, there is a $\gamma \leq \omega_1$ such that $\alpha_n < \gamma$ for all n. Then $\{X_n \mid n < \omega\} \subseteq B_\gamma$. Hence $\bigcup_{n=1}^{\infty} X_n$, $\bigcap_{n=1}^{\infty} X_n \in B_{\gamma+1} \subseteq B$.

The proof is complete. □

We call $\langle B_\alpha \mid \alpha < \omega_1 \rangle$ the *Borel hierarchy*. (Actually, one can define various "Borel hierarchies". We have just chosen a simple one.) If we define the *rank* of a Borel set, X, to be the least γ with $X \in B_{\gamma+1}$, then the rank of a Borel set is a measure of its "complexity". It tells us how many steps it needs to reach X starting with intervals and building up using, at each stage, either a countable union, a countable intersection, or a difference.

Using the Borel hierarchy, we can prove :

1.2 <u>Theorem.</u>

The set of Borel sets has cardinality $2^{\aleph_0}$ (= $|\mathbb{R}|$).

Proof : Since there are $2^{\aleph_0}$ many open intervals (a, b), $|B| \geq 2^{\aleph_0}$. We show that $|B| \leq 2^{\aleph_0}$. We prove first, by induction on α, that $|B_\alpha| \leq 2^{\aleph_0}$ for all $\alpha < \omega_1$.

For $\alpha = 0$, we clearly have $|B_0| = 2^{\aleph_0}$. Now suppose that $|B_\beta| \leq 2^{\aleph_0}$ for all $\beta < \alpha$, where $\lim(\alpha)$, $\alpha < \omega_1$. Then,

$$|B_\alpha| = |\bigcup_{\beta<\alpha} B_\beta| \leq \sum_{\beta<\alpha} |B_\beta| \leq |\alpha|.2^{\aleph_0} = \aleph_0.2^{\aleph_0} = 2^{\aleph_0}.$$

Finally, suppose $|B_\alpha| \leq 2^{\aleph_0}$. Clearly,

$$B_{\alpha+1} = \{X - Y \mid (X, Y) \in B_\alpha \times B_\alpha\}$$
$$\cup\{\cup f[\omega] \mid f : \omega \to B_\alpha\}$$
$$\cup\{\cap f[\omega] \mid f : \omega \to B_\alpha\}.$$

(This is just a formal representation of the definition of $B_{\alpha+1}$ from B_α.) Hence

$$\begin{aligned} |B_{\alpha+1}| &\leq |B_\alpha \times B_\alpha| + |{}^{\omega}B_\alpha| + |{}^{\omega}B_\alpha| \\ &= |B_\alpha|.|B_\alpha| + |B_\alpha|^{\aleph_0} + |B_\alpha|^{\aleph_0} \\ &\leq 2^{\aleph_0}.2^{\aleph_0} + (2^{\aleph_0})^{\aleph_0} + (2^{\aleph_0})^{\aleph_0} \\ &= 2^{\aleph_0}, \text{ as required .} \end{aligned}$$

Thus $|B_\alpha| \leq 2^{\aleph_0}$ for all $\alpha < \omega_1$. Hence

$$|B| = |\bigcup_{\alpha<\omega_1} B_\alpha| \leq \sum_{\alpha<\omega_1} |B_\alpha| \leq \aleph_1.2^{\aleph_0} = 2^{\aleph_0}. \quad \square$$

2. Closed Unbounded Sets

Let λ be a limit ordinal. A set $C \subseteq \lambda$ is *closed* in λ iff $\cup(C \cap \alpha) \in C$ for every limit ordinal $\alpha < \lambda$. That is, C is closed iff, whenever s is an increasing sequence of ordinals which is bounded in C, and whose domain is a limit ordinal, then $\lim(s) \in C$. (Readers who have done Problems I.3 and III.3 will have already met this notion from a topological viewpoint. A set $C \subseteq \lambda$ is closed in the above sense iff it is closed in the order topology on λ.)

Our first result is immediate (even if we do not assume the topological characterisation mentioned above).

2.1 Theorem

If A, B are closed subsets of λ, so too is $A \cap B$. □

A subset C of λ which is at the same time closed and unbounded in λ is said to be *club* in λ. Now, if $cf(\lambda) = \omega$, any ω-sequence cofinal in λ determines a club set (its range). But if $cf(\lambda) > \omega$, any club set in λ is "large", in a sense made precise by the following result.

2.2 Theorem

Suppose $cf(\lambda) > \omega$. If A, B are club in λ, so too is $A \cap B$.

Proof : By virtue of 2.1, we need only prove that $A \cap B$ is unbounded in λ. Let $\alpha \in \lambda$ be given. We seek a $\gamma \in A \cap B$, $\gamma > \alpha$. Choose $\alpha_o \in A$, $\alpha_o > \alpha$. Since A is unbounded, this is always possible. Now choose $\alpha_1 \in B$, $\alpha_1 > \alpha_o$. Since B is unbounded, this is also possible. By recursion now, we can define a sequence $\langle\alpha_n \mid n < \omega\rangle$ so that $\alpha_{n+1} > \alpha_n$ and $\alpha_{2n} \in A$, $\alpha_{2n+1} \in B$. Let $\gamma = \lim_{n<\omega} \alpha_n$. Since $cf(\lambda) > \omega$, $\gamma \in \lambda$. Clearly, $\gamma = \lim_{n<\omega} \alpha_{2n}$, so as $\alpha_{2n} \in A$ for all n and A is closed, $\gamma \in A$. Similarly, $\gamma = \lim_{n<\omega} \alpha_{2n+1}$, so $\gamma \in B$. Hence $\gamma \in A \cap B$ and we are done. □

The next result relates club sets to our old friend the normal function, at least in the case of ω_1.

2.3 Theorem

A set $C \subseteq \omega_1$ is club in ω_1 iff it is the range of a normal function $f : \omega_1 \to \omega_1$.

Proof : Let $C \subseteq \omega_1$ be club in ω_1. By recursion, define a function $f : \omega_1 \to \omega_1$ so that :

$$f(0) = \text{the smallest member of } C \;;$$

$$f(\alpha) = \text{the smallest member of } C - f[\alpha].$$

Since C is unbounded in ω_1 and ω_1 is regular, $|C| = \aleph_1$. Hence f is well-defined. And f is clearly order-preserving. Suppose $\lim(\alpha)$, $\alpha < \omega_1$. Since $f[\alpha] \subseteq C$ and C is closed in ω_1, $\cup f[\alpha] \in C$. Thus, by definition, $f(\alpha) = \cup f[\alpha]$. By III.4.1, therefore, $f(\alpha) = \lim_{\beta<\alpha} f(\beta)$. Hence f is a normal function.

Conversely, if $f : \omega_1 \to \omega_1$ is a normal function, $C = f[\omega_1]$ is clearly club in ω_1. □

In fact, the above theorem generalises immediately from ω_1 to any regular, uncountable cardinal. The same is true of all the results we present in this section, but for definiteness we shall concentrate only on ω_1 from now on. (Besides providing a "concrete" example, the proofs are marginally simpler for the case ω_1, though in all cases the general proof is essentially the same.)

Connected with 2.3 is our next result.

2.4 Theorem

Let $f : \omega_1 \to \omega_1$ be a normal function. The set $C = \{\alpha \in \omega_1 \mid f(\alpha) = \alpha\}$ is club in ω_1.

Proof : That C is closed is immediate. And by III.8.9, C is unbounded in ω_1. □

The following result is often useful.

2.5 Theorem

Let $f : \omega_1 \to \omega_1$. Let $C = \{\alpha \mid f[\alpha] \subseteq \alpha\}$. Then C is club in ω_1.

Proof : Since $f[\alpha] = \bigcup_{\beta<\alpha} f[\beta]$ for any limit ordinal α, it is easily seen that C is closed in ω_1. We prove unboundedness. Let $\alpha_o \in \omega_1$ be given. By recursion, we define an increasing sequence $\langle\alpha_n \mid n < \omega\rangle$ of countable ordinals such that $f[\alpha_n] \subseteq \alpha_{n+1}$ (e.g. with α_{n+1} the least such in each case). Let $\alpha = \lim_{n<\omega} \alpha_n$. Then $f[\alpha] = \bigcup_{n<\omega} f[\alpha_n] \subseteq \bigcup_{n<\omega} \alpha_{n+1} = \alpha$, so $\alpha \in C$. □

We now strengthen 2.2 considerably.

2.6 Theorem

Let A_n, $n = 1, 2, \ldots$ be club subsets of ω_1. Then $A = \bigcap_{n=1}^{\infty} A_n$ is club in ω_1.

Proof : That A is closed is immediate. We prove unboundedness. For each n, let $B_n = \bigcap_{i=1}^{n} A_i$. By 2.2 and induction, each set B_n is club in ω_1. Clearly, $B_1 \supseteq B_2 \supseteq B_3 \supseteq \ldots$, and $A = \bigcap_{n=1}^{\infty} B_n$. Let $\alpha_o \in \omega_1$ be given. Since each B_n is unbounded, we can use the recursion principle to define a sequence $\langle\alpha_n \mid n < \omega\rangle$ such that

$$\alpha_{n+1} \text{ is the least ordinal in } B_{n+1} - (\alpha_n + 1).$$

Let $\alpha = \lim_{n<\omega} \alpha_n$. For each n, $\alpha = \lim_{m<\omega} \alpha_{n+m}$, so $\alpha \in B_n$. Hence $\alpha \in A$, as required. □

3. Stationary Sets and Regressive Functions

This section depends upon section 2. As there, although we concentrate on ω_1, all of our results hold for any uncountable regular cardinal, the proofs differing only slightly from the proofs for ω_1.

A set $E \subseteq \omega_1$ is said to be *stationary* iff $E \cap C \neq \emptyset$ for every club set $C \subseteq \omega_1$. By 2.2, every club set is stationary. The converse fails trivially. For example, let $E = \omega_1 - \{\omega\}$. E is not closed, since the limit of $\langle n \mid n < \omega \rangle$ does not lie in E. But E is stationary. For if $C \subseteq \omega_1$ is club, then by the unboundedness of C alone, $E \cap C \neq \emptyset$. Nevertheless, stationary sets are "large". They are certainly unbounded. (To see this, observe that each set $C_\alpha = \omega_1 - \alpha$ is club, for $\alpha < \omega_1$.) The following result, though a trivial consequence of the definition, is worth stating as a lemma :

3.1 Lemma

A set $E \subseteq \omega_1$ is stationary iff $\omega_1 - E$ does not contain a club set. □

On ω_1, it is not easy to *find* examples of stationary sets which are not already club or else simple modifications of club sets. On ω_2 one can obtain a pair of good examples : the sets

$$\{ \alpha \in \omega_2 \mid \lim(\alpha) \ \& \ \mathrm{cf}(\alpha) = \omega \}$$

$$\{ \alpha \in \omega_2 \mid \lim(\alpha) \ \& \ \mathrm{cf}(\alpha) = \omega_1 \}$$

form a pair of disjoint stationary sets. However, our inability to *find* nice examples on ω_1 does not mean they do not exist. Indeed, the following classical result (which we do not prove here) tells us that there are stationary subsets of ω_1 which do not at all resemble club sets.

3.2 Theorem

If $E \subseteq \omega_1$ is stationary, there are stationary sets A_α, $\alpha < \omega_1$, such that :

(i) $\alpha \neq \beta \rightarrow A_\alpha \cap A_\beta = \emptyset$;

(ii) $\bigcup_{\alpha<\omega_1} A_\alpha = E$. □

We shall obtain a surprising and very useful characterisation of stationary sets in terms of "regressive functions" on ω_1.

A function $f : \omega_1 \to \omega_1$ is *regressive* iff $f(\alpha) < \alpha$ for every non-zero α in ω_1. Similarly, if $E \subseteq \omega_1$, a function $f : E \to \omega_1$ is *regressive* iff $f(\alpha) < \alpha$ for every non-zero α in E.

In order to prove our desired equivalence (not yet stated), we need a result on club sets which extends 2.6.

Let $\langle C_\alpha \mid \alpha < \omega_1\rangle$ be an ω_1-sequence of club sets. Now, it is not necessarily the case that $\bigcap_{\alpha<\omega_1} C_\alpha$ is club. Indeed, it may be empty (let $C_\alpha = \omega_1 - \alpha$, say). The *diagonal intersection* of this sequence is the set

$$\Delta_{\alpha<\omega_1} C_\alpha = \{\gamma \in \omega_1 \mid (\forall\alpha < \gamma)(\gamma \in C_\alpha)\} .$$

Thus, $\gamma \in \Delta_{\alpha<\omega_1} C_\alpha$ iff $\gamma \in \bigcap_{\alpha<\gamma} C_\alpha$.

Clearly, $\bigcap_{\alpha<\omega_1} C_\alpha \subseteq \Delta_{\alpha<\omega_1} C_\alpha$. But the two "intersections" are quite different, as the next result shows.

3.3 Theorem

If $\langle C_\alpha \mid \alpha < \omega_1\rangle$ is a sequence of club sets in ω_1, then $\Delta_{\alpha<\omega_1} C_\alpha$ is club in ω_1.

Proof : Let $C = \Delta_{\alpha<\omega_1} C_\alpha$. We show that if γ is a countable limit ordinal, then $\bigcup(C \cap \gamma) \in C$. If $C \cap \gamma$ is bounded in γ, then

$$\bigcup(C \cap \gamma) = \max(C \cap \gamma) \in C,$$

so we are done. So assume $C \cap \gamma$ is unbounded in γ. Let $\langle\gamma_n \mid n < \omega\rangle$ be a strictly increasing sequence of elements of C cofinal in γ. For each n, $\{\gamma_n, \gamma_{n+1}, \gamma_{n+2}, \ldots\} \subseteq \bigcap_{\alpha<\gamma_n} C_\alpha$. Hence, as $\bigcap_{\alpha<\gamma_n} C_\alpha$ is club (by 2.6), $\gamma = \lim_{m<\omega} \gamma_{n+m} \in \bigcap_{\alpha<\gamma_n} C_\alpha$. But n was arbitrary here. So,

$$\gamma \in \bigcap_{n<\omega} \bigcap_{\alpha<\gamma_n} C_\alpha = \bigcap_{\alpha<\gamma} C_\alpha .$$

Thus $\gamma \in C$. Since $\gamma = \cup(C \cap \gamma)$ in this case, we are done.

That proves that C is closed in ω_1. We check unboundedness now. Let $\alpha_o \in \omega_1$ be given. Since $\bigcap_{\alpha<\alpha_o} C_\alpha$ is club (by 2.6), we can find an $\alpha_1 \in \omega_1$, $\alpha_1 > \alpha_o$, so that $\alpha_1 \in \bigcap_{\alpha<\alpha_o} C_\alpha$. In this manner, we may use the recursion principle to define a sequence $\langle\alpha_n \mid n < \omega\rangle$ so that $\alpha_{n+1} > \alpha_n$ and $\alpha_{n+1} \in \bigcap_{\alpha<\alpha_n} C_\alpha$. Let $\gamma = \lim_{n<\omega} \alpha_n$. Since $\{\alpha_n, \alpha_{n+1}, \alpha_{n+2}, \ldots\} \subseteq \bigcap_{\alpha<\alpha_n} C_\alpha$ for each n and $\gamma = \lim_{m<\omega} \alpha_{n+m}$, $\gamma \in \bigcap_{\alpha<\alpha_n} C_\alpha$ for each n, so $\gamma \in \bigcap_{n<\omega} \bigcap_{\alpha<\alpha_n} C_\alpha = \bigcap_{\alpha<\gamma} C_\alpha$, which means $\gamma \in C$. Since $\gamma > \alpha_o$, we are done. □

3.4 Theorem

Let $E \subseteq \omega_1$. The following are equivalent:

(i) E is stationary ;

(ii) If $f : E \to \omega_1$ is regressive, then for some $\gamma \in \omega_1$, $f^{-1}[\gamma]$ is stationary in ω_1 ;

(iii) If $f : E \to \omega_1$ is regressive, then for some $\gamma \in \omega_1$, $f^{-1}[\gamma]$ is unbounded in ω_1 .

Proof : (i) → (ii). Suppose that (i) holds but, contrary to (ii) there is a regressive function $f : E \to \omega_1$ such that for no $\gamma \in \omega_1$ is $f^{-1}[\gamma]$ stationary. Thus for each $\gamma \in \omega_1$ we can find a club set $C_\gamma \subseteq \omega_1$ such that $f^{-1}[\gamma] \cap C_\gamma = \emptyset$. Let $C = \Delta_{\gamma<\omega_1} C_\gamma$. We prove that $C \cap E = \emptyset$, contradicting (i). (This uses 3.3, of course.) Well, suppose otherwise, and let $\alpha \in C \cap E$. Since $\alpha \in C$, $\alpha \in \bigcap_{\gamma<\alpha} C_\gamma$.

Since $\alpha \in E$, $\gamma = f(\alpha) < \alpha$ is defined. Thus $\alpha \in f^{-1}[\gamma]$. Thus $\alpha \notin C_\gamma$. Hence $\alpha \notin \bigcap_{\gamma<\alpha} C_\gamma$, a contradiction. This proves that $C \cap E = \emptyset$, and completes the proof that (i) implies (ii).

(ii) $\rightarrow$ (iii). Trivial.

(iii) $\rightarrow$ (i). Assume $\neg$(i). Let $C \subseteq \omega_1$ be club with $C \cap E = \emptyset$. Define $f : E \rightarrow \omega_1$ by letting $f(\alpha) = \cup(C \cap \alpha)$. Since C is club and $C \cap E = \emptyset$, $f(\alpha) = \max(C \cap \alpha) < \alpha$ for all non-zero α in E. That is, f is regressive. Let $\gamma \in \omega_1$. Since C is unbounded in ω_1 we can pick $\alpha \in C$, $\alpha > \gamma$. If now $\delta \in E$ is such that $\delta > \alpha$, then $f(\delta) \geq \alpha$, so $f(\delta) \neq \gamma$. Hence $f^{-1}[\gamma] \subseteq \alpha+1$. Since γ was arbitrary, this proves $\neg$(iii) (for this f). $\square$

4. Trees

A *tree* is a poset $\underset{\sim}{T} = (T_1 \leq_T)$ such that for every $x \in T$, the set

$$\hat{x} = \{y \in T \mid y <_T x\}$$

of all predecessors of x is well-ordered by $<_T$. The ordinal number $\mathrm{Ord}(\hat{x}, <_T)$ is called the *height* of x in $\underset{\sim}{T}$, $\mathrm{ht}(x)$. This provides us with a highly intuitive (but sometimes misleading) picture of a tree:

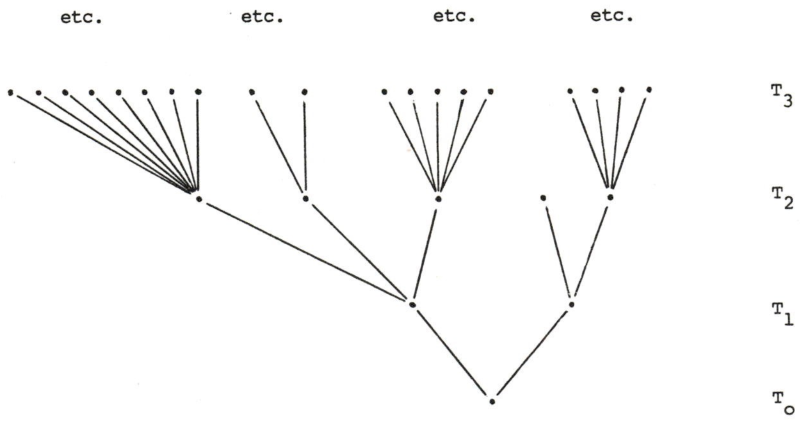

The elements (or *nodes*) of the tree are denoted by points, and a vertical line drawn between two points indicates that the higher point succeeds the lower in the ordering $<_T$. If we set

$$T_\alpha = \{ x \in T \mid ht(x) = \alpha \}$$

for each ordinal α, we obtain a stratification of $\underset{\sim}{T}$ into *levels*: T_α is the α'th *level* of $\underset{\sim}{T}$. Any element of T_α will have exactly α predecessors in $\underset{\sim}{T}$. (Consequently no two elements of the same level will be comparable in $<_T$.) Notice that if we follow a "path" through the tree moving upwards, each choice of direction upon leaving a node is irrevocable: no two paths ever coincide once they have split up. Or, to put it another way, if we are at any node in the tree, there is one and only one route of descent through the tree.

Clearly, if $\underset{\sim}{T}$ is a tree and if $T_\alpha \neq \emptyset$, then $T_\beta \neq \emptyset$ for all $\beta < \alpha$. (Each of the α many predecessors of any element of T_α will have to lie in each of the levels T_β for $\beta < \alpha$.) Now, (as T must be a set) there is a unique least λ such that $T_\lambda = \emptyset$; and it follows from our last remark that $T_\alpha = \emptyset$ for all $\alpha \geq \lambda$. We call λ the *height* of $\underset{\sim}{T}$, $ht(\underset{\sim}{T})$.

A *chain* in $\underset{\sim}{T}$ is a linearly ordered subset of $\underset{\sim}{T}$; a *branch* is a chain which is closed under predecessors. For example, for any x in $\underset{\sim}{T}$, $\hat{x}$ is a branch in $\underset{\sim}{T}$. But branches do not need to be of this form : they may have order type ω and have no last element. One of the most basic, and useful, questions one can ask about trees is : Under what conditions does a tree have an infinite branch? (We shall see an application of this when we have proved our next theorem. The example we give is of a very simple type. We reprove Cantor's theorem that there are uncountably many reals. To present a deeper example would require too great a digression.)

4.1 <u>Theorem</u> (König Tree Lemma)

Let $\underset{\sim}{T} = (T, \leq_T)$ be a tree of height ω such that every level of $\underset{\sim}{T}$ is finite.

Then $\underset{\sim}{T}$ has an infinite branch.

Proof : For each $x \in T$, let $[x]$ denote the set of all successors of x in $\underset{\sim}{T}$,

$$[x] = \{ y \in T \mid x <_T y \} .$$

Clearly, the sets $[x]$, $x \in T_0$, constitute a (disjoint) partition of $T - T_0$ into finitely many pieces. Since T is infinite and T_0 is finite, $T - T_0$ is infinite. Hence as $T - T_0 = \bigcup_{x \in T_0} [x]$, for at least one $x \in T_0$, $[x]$ must be infinite. Let x_0 be such an x. Again, $[x_0] - T_1$ is infinite, and the sets $[x]$, $x \in T_1 \cap [x_0]$, partition $[x_0] - T_1$ into finitely many pieces so we can pick $x_1 \in T_1 \cap [x_0]$ so that $[x_1]$ is infinite. Proceeding in this manner (more formally, by appealing to the recursion principle) we can define a sequence $\langle x_n \mid n < \omega \rangle$ so that $x_{n+1} \in T_{n+1} \cap [x_n]$ and $[x_{n+1}]$ is infinite. Clearly, $\{x_n \mid n < \omega\}$ is an infinite branch of $\underset{\sim}{T}$. □

4.2 <u>Corollary</u>

There are uncountably many reals.

Proof : What we show is that there are uncountably many members of the set ${}^{\omega}2$. (By considering binary representations of reals in $(0, 1)$, this is easily seen to yield the desired result.)

Suppose otherwise. That is, suppose $2^{\aleph_0} = \aleph_0$. Let $\langle \varepsilon_n \mid n < \omega \rangle$ enumerate ${}^{\omega}2$. Let

$$T_n = \{ \varepsilon \in {}^{n}2 \mid (\forall m < n)(\varepsilon \restriction m+1 \neq \varepsilon_m \restriction m+1) \} .$$

Ordered by inclusion (i.e. functional extension), $T = \bigcup_{n<\omega} T_n$ is a tree. (This is easily seen). It is easily seen to be infinite, and clearly has height at most ω. Since $|{}^{n}2| = 2^n < \aleph_0$, each level is finite, so T does have height ω. Let b be an infinite branch. Set $f = \bigcup b$. Clearly, $f \in {}^{\omega}2$. But for all n, $f \restriction n+1 \in T$, so $f\upharpoonright n+1 \neq \varepsilon_n\upharpoonright n+1$. Thus $f \notin \{\varepsilon_n \mid n < \omega\}$, a contradiction. □

Of course, in the above example one gains nothing by using the König Lemma. Indeed, the proof that T is infinite essentially involves the classical Cantor proof of 4.2. However, it does illustrate how one can view inductive procedures of the Cantor type as applications of the König lemma. And this has an advantage in more complex situations of the type discussed in section 5.

And now to a cautionary tale. What is your immediate response to the following question?

> Let $\underset{\sim}{T}$ be a tree of height ω_1, all of whose levels are countable. Does $\underset{\sim}{T}$ necessarily have an uncountable branch?

At first glance, one would think that the proof of 4.1 will generalise easily to give a positive answer to this question. And rare is the beginner who sees at once that this is not the case.

In a moment we shall produce an example of a tree of height ω_1, having only countable levels, with no uncountable branch. (Indeed, our tree will have a stronger property : each member of the tree will have extensions on arbitrarily high levels of the tree! But still there will be no uncountable branch.) First let us try to see what can possibly go wrong when we try to generalise the proof of 4.1.

We pick $x_o \in T_o$ to have uncountably many extensions. Then we pick $x_1 \in T_1$ to extend x_o and itself have uncountably many extensions. And so on. This is alright so far. But now we have a branch $\{x_n \mid n < \omega\}$. We want to pick a point $x_\omega \in T_\omega$ so that $x_n <_T x_\omega$ for all n. But there need not necessarily be such an x_ω! The branch $\{x_n \mid n < \omega\}$ may not extend onto T_ω. Of course, your reaction is that if we are careful in how we choose x_1, x_2, etc, we can ensure that we do obtain a branch which extends. True enough. But there is a long way to go to ω_1. And we cannot allow *in advance* for all future limit levels. Sooner or later we will reach a limit stage which we have not been able to allow for, and then the same problem arises. Of course, our reader is still unconvinced (?). So without any more ado, let us convince him in the only sure manner. We construct our counterexample. (It is due to *N. Aronszajn.*)

4.3 <u>Theorem</u>

There is a tree $\underset{\sim}{T} = (T, \leq_T)$ such that:

(i) $\underset{\sim}{T}$ has height ω_1 ;

(ii) $|T_\alpha| \leq \aleph_0$ for all $\alpha < \omega_1$;

(iii) if $x \in T_\alpha$ and $\alpha < \beta < \omega_1$, there is a $y \in T_\beta$ such that $x <_T y$;

(iv) $\underset{\sim}{T}$ has no uncountable branch.

Proof : The elements of T_α will be strictly increasing α-sequences of rational numbers which are bounded above. The ordering of $\underset{\sim}{T}$ will be inclusion (i.e. sequence extension). Notice at once that this will yield condition (iv) of the theorem. For an uncountable branch of our tree would present us with a strictly increasing ω_1-sequence of rationals, which is impossible. Notice also that condition (i) will follow from condition (iii). So what we must do is construct our tree to satisfy both (ii) and (iii). This requires some care.

The definition is by recursion on the levels. That is, we define T_α from $\bigcup_{\beta<\alpha} T_\beta$. (Since the ordering is inclusion, we are only concerned with *which* sequences each T_α will contain.) We use $T \upharpoonright \alpha$ to denote the set $\bigcup_{\beta<\alpha} T_\beta$.

The recursion is carried out so as to preserve the following condition:

(*) If $s \in T_\alpha$ and $\alpha < \beta < \omega_1$, then for each rational number $q > \sup(s)$ there is a $t \in T_\beta$ such that $s \subset t$ and $\sup(t) < q$.

To commence we set $T_0 = \{\emptyset\}$. If $T \upharpoonright \alpha+1$ is defined, $T_{\alpha+1}$ is defined as

$$T_{\alpha+1} = \{s \in {}^{\alpha+1}\mathbb{Q} \mid s \restriction \alpha \in T_\alpha\}.$$

If $|T_\alpha| \leqslant \aleph_0$, then since $|\mathbb{Q}| = \aleph_0$, we have $|T_{\alpha+1}| = \aleph_0$. Moreover, if (*) is valid for $T \restriction \alpha+1$ it will clearly be valid for $T \restriction \alpha+2$.

There remains the case when $T \restriction \alpha$ is defined for α a limit ordinal. Let us call a branch b of $T \restriction \alpha$ *cofinal* if it intersects each level of $T \restriction \alpha$ (i.e. if its order type under the tree ordering is α). In order to define T_α, we must extend some cofinal branches of $T \restriction \alpha$. Indeed, any element of T_α will necessarily be of the form $\cup b$, where b is a cofinal branch of $T \restriction \alpha$. Now, if $\cup b \in T_\alpha$, $\cup b$ must be bounded above in $\mathbb{Q}$, so it must be the use that $\{\sup(s) \mid s \in b\}$ is bounded above in $\mathbb{Q}$. It is in order to ensure that such branches b can always be found that we introduced the requirement (*). (This will become clear in a moment.) Now we cannot simply extend *all* such branches, since there may be uncountably many of them, which would make T_α uncountable. On the other hand, we must ensure that (*) holds for $T \restriction \alpha+1$. So we proceed as follows.

Notice first that (*) will hold for $T \restriction \alpha$ providing it holds for each $T \restriction \beta$, $\beta < \alpha$. Let $\langle \alpha_n \mid n < \omega \rangle$ be a strictly increasing sequence of ordinals cofinal in α. For each $s \in T \restriction \alpha$ and each rational number $q > \sup(s)$ we define an element $b(s,q)$ of ${}^\alpha\mathbb{Q}$ as follows. Let $n(s)$ be least with $s \in T \restriction \alpha_{n(s)}$. By (*), pick $s_{n(s)} \in T_{\alpha_{n(s)}}$ so that $s \subset s_{n(s)}$ and $\sup(s_{n(s)}) < q$. We define s_n for $n(s) < n < \omega$ now by recursion. Let $s_{n+1} \in T_{\alpha_{n+1}}$ be such that $s_n \subset s_{n+1}$ and $\sup(s_{n+1}) < q$. If $\sup(s_n) < q$, then by (*) such an s_{n+1} can always be found. Let

$$b(s,q) = \bigcup_{n=n(s)}^{\infty} s_n .$$

Clearly, $b(s,q) \in {}^\alpha\mathbb{Q}$ and $s \subset b(s,q)$. Moreover, $\sup b(s,q) \leqslant q$. We set

$$T_\alpha = \{b(s,q) \mid s \in T \restriction \alpha \ \&\ q \in \mathbb{Q} \ \&\ q > \sup(s)\}.$$

If $|T \restriction \alpha| \leqslant \aleph_0$, then $|T_\alpha| \leqslant \aleph_0$, and $T \restriction \alpha+1$ satisfies (*) by virtue of the construction.

That completes the description of the definition of T. An easy induction on the levels shows that condition (ii) holds. (The induction steps have already been noted.) And (iii) follows directly from (*). The proof is complete. □

5. Extensions of Lebesgue Measure

Let us commence by recalling some standard definitions from measure theory.

Let $\mathcal{A}$ be a σ-field of subsets of X. (See Problem I.1 and the introduction to section 1 for the relevant definitions.) A *measure* on $\mathcal{A}$ is a function $\mu : \mathcal{A} \to [0, 1]$ (the closed unit interval) such that:

(i) $\mu(\emptyset) = 0, \quad \mu(X) = 1$;

(ii) if $\{E_n\}$ is a finite or infinite sequence of disjoint elements of $\mathcal{A}$, then $\mu(\cup_n E_n) = \sum_n \mu(E_n)$.

The classic example of such is where $X = [0, 1]$, $\mathcal{L}$ is the σ-field of all Lebesgue measurable subsets of X, and μ is the Lebesgue measure on $\mathcal{L}$. Now, it is known that $\mathcal{L} \neq \mathcal{P}(X)$. (The proof uses the Axiom of Choice in a fairly strong form : without such an assumption the result need not be valid.) A natural question to ask is whether it is possible to extend the Lebesgue measure on $\mathcal{L}$ to a measure defined on the whole σ-field $\mathcal{P}(X)$; i.e. does there exist an extension of Lebesgue measure defined on all sets of reals (in [0, 1])? The usual proof that $\mathcal{L} \neq \mathcal{P}(X)$ adapts easily to a proof that such an extension would fail to be translation invariant, but does not tell us anything more. In fact the question has a rather unexpected consequence : we devote this section to a proof of the following result :

5.1 Theorem

Assume there is an extension of Lebesgue measure defined on all sets of reals (in [0, 1]). Then there is a weakly inaccessible cardinal $\kappa \leq 2^{\aleph_0}$. □

As an immediate consequence of 5.1 we have :

5.2 Corollary

Assume CH. Then there is no extension of Lebesgue measure defined on all sets of reals.

Proof : $2^{\aleph_0}$ cannot at the same time equal $\aleph_1$ and dominate a weakly inaccessible cardinal. □

For the rest of this section we shall denote $2^{\aleph_0}$ by λ. We shall assume that there is an extension of Lebesgue measure to a measure defined on all sets of reals in $[0, 1]$. (In fact, the proof works for any measure : it need not extend Lebesgue measure.) Since $|[0, 1]| = \lambda$, the measure on $\mathcal{P}([0, 1])$ induces a measure on $\mathcal{P}(\lambda)$ in a trivial manner. If μ is a measure defined on $\mathcal{P}(\lambda)$ we shall simply say that μ is a *measure on* λ. It is with measures on cardinals that we shall be concerned.

If μ is any measure on some uncountable cardinal κ, and if θ is also an uncountable cardinal, we shall say that μ is θ-*additive* iff, whenever $\psi < \theta$ and E_ν, $\nu < \psi$, are sets of measure zero, then $\bigcup_{\nu<\psi} E_\nu$ has measure zero. (Thus, by definition, any measure is $\aleph_1$-additive.)

Suppose μ is a measure on the uncountable cardinal κ. There is clearly a largest cardinal θ such that μ is θ-additive. Now clearly, $\theta \geq \aleph_1$. And since $\kappa = \bigcup_{\alpha<\kappa} \{\alpha\}$, we also have $\theta \leq \kappa$. By the definition of θ, there is a set A of positive measure for which there are disjoint sets A_ν, $\nu < \theta$, of measure zero with $A = \bigcup_{\nu<\theta} A_\nu$. Now define a map $f : A \to \theta$ by

$$f(a) = \nu \quad \leftrightarrow \quad a \in A_\nu .$$

And for $B \subseteq \theta$, set

$$\sigma(B) = \frac{\mu(f^{-1}[B])}{\mu(A)} .$$

It is easily seen that σ is a θ-additive measure on θ. Hence, if we use the word *strong* to describe a measure on a cardinal κ which is κ-additive, we see that we have proved the following lemma

5.3 Lemma

Let κ be an uncountable cardinal, μ a measure on κ. Then there is an uncountable cardinal $\theta \leq \kappa$ such that there is a strong measure on θ. □

Now, we are assuming that there is a measure on λ. So by 5.3 we conclude that there is an uncountable cardinal $\kappa \leq \lambda$ and a strong measure μ on κ. We fix κ and μ from now on. We complete the proof of 5.1 by showing that κ is weakly inaccessible. In order to show that κ is regular we require a simple lemma.

5.4 Lemma

If $\xi < \kappa$, then $\xi = \{ \alpha \mid \alpha < \xi \}$ has measure zero.

Proof : Since μ is strong (i.e. κ-additive.) □

5.5 Lemma

κ is regular.

Proof : Suppose not. Then there is a $\theta < \kappa$ and ordinals $\kappa_\nu < \kappa$, $\nu < \theta$, with $\kappa = \bigcup_{\nu<\theta} \kappa_\nu$. By 5.4, $\mu(\kappa_\nu) = 0$ for all ν. So, as μ is strong, $\mu(\kappa) = 0$, which is impossible. □

The proof that κ is a limit cardinal will take a little longer.

We say that μ is *normal* iff, whenever $B \subseteq \kappa$ has positive measure and $f : B \to \kappa$ is such that $f(\xi) < \xi$ for all $\xi \in B$, there is a $B' \subseteq B$ of positive measure such that f is constant on B'. As a first step in our proof we show that we may assume that our measure μ is normal. We require some auxiliary definitions.

Let $f : A \to \kappa$, where $A \subseteq \kappa$. We say that f is *almost bounded* if there is a $\lambda < \kappa$ such that $\{ \xi \in A \mid f(\xi) > \lambda \}$ has measure zero. We say that f is *nowhere bounded* if for each $\lambda < \kappa$, $\{ \xi \in A \mid f(\xi) \leq \lambda \}$ has measure zero. We say that f is *incompressible* if f is nowhere bounded and whenever $B \subseteq A$ has positive measure and $g : B \to \kappa$ and $g(\xi) < f(\xi)$ for all $\xi \in B$, then g is almost bounded. These concepts are only important if A has positive measure, of course, for if A has measure zero, then any map f is simultaneously almost bounded, nowhere bounded, and incompressible.

5.6 Lemma

Let $f : A \to \kappa$ be nowhere bounded. Then we can write A as the disjoint union of sets B and C such that :

(1) $f \upharpoonright B$ is incompressible ;

(2) there is a $g : C \to \kappa$ such that $g(\xi) < f(\xi)$ for all $\xi \in C$ and such that g is nowhere bounded.

Proof : Using Zorn's lemma we obtain a maximal family

$$\mathfrak{F} = \{ (C_i, g_i) \mid i \in K \}$$

such that :

(1) $C_i \subseteq A$ has positive measure ;

(2) $g_i : C_i \to \kappa$ is nowhere bounded ;

(3) $\xi \in C_i \to g_i(\xi) < f(\xi)$;

(4) if i, j are distinct elements of K, then $C_i \cap C_j = \emptyset$.

By (1) and (4), K must be countable. (If K were uncountable, then for some positive number n, C_i would have measure greater than 1/n for uncountably many i, contrary to the measure being finite.) We set

$$C = \cup_{i\in K} C_i , \quad g = \cup_{i\in K} g_i .$$

Since K is countable, g is nowhere bounded. And clearly, $\xi \in C \to g(\xi) < f(\xi)$. Set $B = A - C$. If B has measure zero, $f \restriction B$ is trivially incompressible. If B has positive measure, the maximality of $\mathcal{F}$ implies that $f \restriction B$ is incompressible. □

5.7 Lemma

There is an incompressible function $f : \kappa \to \kappa$.

Proof : We define a sequence of sets $\{A_n\}$ and functions $h_n : A_n \to \kappa$ by induction on n. To commence we set $A_o = \kappa$ and let h_o be the identity function on κ. (By 5.4, h_o is nowhere bounded.) Suppose now that $n = k + 1$ and that we have defined A_k and h_k with $h_k : A_k \to \kappa$ nowhere bounded. We apply 5.6 to A_k, h_k to obtain a set $A_{k+1} \subseteq A_k$ and a map $h_{k+1} : A_{k+1} \to \kappa$ which is nowhere bounded, such that $h_{k+1}(\xi) < h_k(\xi)$ for $\xi \in A_{k+1}$ and $h_k \restriction A_k - A_{k+1}$ is incompressible.

We show that $\bigcap_{n=1}^{\infty} A_n = \emptyset$. Suppose otherwise, and let ξ lie in this intersection. Then $h_o(\xi) > h_1(\xi) > \ldots$, so $\{h_n(\xi)\}$ is a strictly decreasing sequence of ordinals. But the ordinals are well-ordered, so this is impossible.

By the above,

$$\kappa = \bigcup_{n=1}^{\infty} (A_n - A_{n+1})$$

is a disjoint union and we may define $h : \kappa \to \kappa$ by

$$h(\xi) = h_n(\xi) \quad \text{iff} \quad \xi \in A_n - A_{n+1}.$$

It is easily checked that h is incompressible. □

Let $h : \kappa \to \kappa$ be incompressible now. Define a function $\nu : \mathcal{P}(\kappa) \to [0, 1]$ by

$$\nu(A) = \mu(h^{-1}[A]).$$

It is easily checked that ν is a strong measure on κ. We prove that ν is normal. Suppose A has positive ν-measure and that $g : A \to \kappa$ is such that $g(\xi) < \xi$ for all $\xi \in A$. Let $B = h^{-1}[A]$. Then B has positive μ-measure. Let $f = g \circ h$. Thus $f : B \to \kappa$. If $\gamma \in B$, $f(\gamma) = g(h(\gamma)) < h(\gamma)$. So as h is incompressible there is a $\lambda < \kappa$ for which $\{\gamma \in B \mid f(\gamma) \leq \lambda\}$ has positive μ-measure. Since μ is strong and $\lambda < \kappa$ there is a $\lambda' \leq \lambda$ such that $D = \{\gamma \in B \mid f(\gamma) = \lambda'\}$ has positive μ-measure. Let $E = \{\gamma \in A \mid g(\gamma) = \lambda'\}$. Then $\gamma \in D \leftrightarrow g(h(\gamma)) = \lambda' \leftrightarrow h(\gamma) \in E$. Thus $D = h^{-1}[E]$. Hence E has positive ν-measure. But $E \subseteq A$ and g is constant on E, so we are done.

We thus see that without loss of generality we may assume that in fact the measure μ is normal.

5.8 Lemma

Let A have positive measure. Let $h : A \to \kappa$ be such that $h(\xi) < \xi$ for all $\xi \in A$. Then h is almost bounded.

Proof : Let

$$E = \{ \lambda < \kappa \mid h^{-1}[\{\lambda\}] \text{ has positive measure } \}.$$

Clearly, E must be countable.

Let
$$B = \{ \gamma \in A \mid h(\gamma) \notin E \}.$$

B must have measure zero. For otherwise, by normality, there is a $\lambda < \kappa$ and a $B' \subseteq B$ of positive measure such that $h(\gamma) = \lambda$ for $\gamma \in B'$, giving $\lambda \in E$, contrary to the definition of B.

Since κ is regular, $\lambda_o = \sup(E) < \kappa$. But

$$\{ \gamma \in A \mid h(\gamma) > \lambda_o \} \subseteq B.$$

Thus h is almost bounded. □

5.9 Lemma

For almost all $\alpha \in \kappa$, α is a regular cardinal.

Proof : Suppose not. Let $E = \{\alpha\in\kappa \mid cf(\alpha) < \alpha\}$. Thus E has positive measure. Hence, by normality, there is a $\lambda < \kappa$ such that $E_1 = \{\alpha\in\kappa \mid cf(\alpha) = \lambda\}$ has positive measure. For each $\alpha \in E_1$, pick a mapping $h_\alpha : \lambda \to \alpha$ such that $\sup(h_\alpha[\lambda]) = \alpha$. Define for each $\xi < \lambda$ a map $g_\xi : E_1 \to \kappa$ by $g_\xi(\alpha) = h_\alpha(\xi)$. Then for all $\alpha \in E_1$, $g_\xi(\alpha) = h_\alpha(\xi) < \alpha$. Applying 5.8 to g_ξ, we see that there is a set N_ξ of measure zero and an ordinal $\gamma_\xi < \kappa$ such that $g_\xi(\alpha) \leq \gamma_\xi$ if $\alpha \in E_1 - N_\xi$. Set $\gamma = \sup\{\gamma_\xi \mid \xi < \lambda\}$. Since κ is regular, $\gamma < \kappa$. Put $E_2 = E_1 - \cup_{\xi<\lambda} N_\xi$. Since μ is strong, $\mu(E_2) > 0$. Now, for $\alpha \in E_2$,

$$\alpha = \sup\{g_\xi(\alpha) \mid \xi < \lambda\} \leq \sup\{\gamma_\xi \mid \xi < \lambda\} \leq \gamma.$$

Thus $E_2 \subseteq \{\alpha \mid \alpha < \gamma\}$. Since $\mu(E_2) > 0$, this contradicts 5.4. □

We are now able to complete the proof of 5.1, by proving that κ is weakly inaccessible. Suppose not. Thus $\kappa = \theta^+$ for some cardinal θ. Then

$$\{ \alpha \in \kappa \mid \alpha \text{ is regular } \} \subseteq \{ \alpha \in \kappa \mid \alpha \leq \theta \}.$$

Thus by 5.4, $\{\alpha \in \kappa \mid \alpha \text{ is regular}\}$ has measure zero. This contradicts 5.9. Hence κ must be weakly inaccessible, which proves 5.1.

6. A Result About the GCH.

We have already indicated in III.10 that the GCH cannot be proved in the Zermelo-Fraenkel set theory. In fact, using techniques of the kind outlined in Chapter VI, it may be shown that for any uncountable regular cardinal κ it is consistent with the ZFC axioms that the GCH holds below κ (i.e. $\lambda < \kappa \to 2^\lambda = \lambda^+$) but fails at κ itself (i.e. $2^\kappa > \kappa^+$). For instance, it is consistent with the ZFC axioms that $2^{\aleph_0} = \aleph_1$, $2^{\aleph_1} = \aleph_2$, $2^{\aleph_2} = \aleph_3$, but $2^{\aleph_3} = \aleph_{17}$. But the regularity of κ is essential here. Or almost: if κ is singular of cofinality ω

it may be possible for the GCH to hold below κ and still fail at κ, but the situation is rather complex. However, if κ is singular of uncountable cofinality, the validity of the GCH below κ implies its validity at κ. The proof of this result, which we present here, is non-trivial, and provides a good illustration of an argument in combinatorial set theory and cardinal arithmetic. The proof requires a knowledge of sections 2 and 3.

We fix from now on a singular cardinal κ of uncountable cofinality. We assume that $2^{\lambda} = \lambda^{+}$ for all $\lambda < \kappa$. We prove that $2^{\kappa} = \kappa^{+}$.

Let $\theta = \mathrm{cf}(\kappa)$. Thus $\omega_1 \le \theta < \kappa$. Let $\langle \kappa_\nu \mid \nu < \theta \rangle$ be a normal sequence of cardinals which is cofinal in κ.

6.1 Lemma

Let $E \subseteq \theta$ be stationary. Let $f : E \to \kappa$ be such that $f(\alpha) < \kappa_\alpha$ for all $\alpha \in E$. Then there is a $\gamma < \theta$ and a stationary set $E' \subseteq E$ such that

$$\alpha \in E' \rightarrow f(\alpha) < \kappa_\gamma.$$

Proof : Let $C = \{\alpha \in \theta \mid \lim(\alpha)\}$. For $\alpha \in C$, $\kappa_\alpha = \lim_{\nu<\alpha} \kappa_\nu$. So, for each $\alpha \in E \cap C$ there is a $\nu < \alpha$ such that $f(\alpha) < \kappa_\nu$. Let $g(\alpha)$ be the least such ν. Clearly, $g : E \cap C \to \theta$ is regressive. But E is stationary in θ and C is club in θ and θ is an uncountable regular cardinal. So, $E \cap C$ is stationary and, by 3.4 there is a stationary set $E' \subseteq E \cap C$ and a $\gamma < \theta$ with

$$\alpha \in E' \rightarrow g(\alpha) = \gamma.$$

Thus $$\alpha \in E' \rightarrow f(\alpha) < \kappa_\gamma . \quad \square$$

Now, by assumption, $2^{\kappa_\alpha} = \kappa_\alpha^{+}$ for each $\alpha < \theta$. Let $\langle A^\alpha_\xi \mid \xi < \kappa_\alpha^{+} \rangle$ be an enumeration of $\mathcal{P}(\kappa_\alpha)$.

For $A \subseteq \kappa$, define $f_A : \theta \to \kappa$ by

$$f_A(\alpha) = \xi \quad \text{iff} \quad A \cap \kappa_\alpha = A^\alpha_\xi.$$

Notice that if $A, B \subseteq \kappa$ and $A \neq B$, then for some $\alpha < \theta$, $A \cap \kappa_\alpha \neq B \cap \kappa_\alpha$, whence $f_A(\beta) \neq f_B(\beta)$ whenever $\beta \geq \alpha$, which means that the set

$$\{ \alpha \in \theta \mid f_A(\alpha) = f_B(\alpha) \}$$

is bounded in θ.

We define a relation R on $\mathcal{P}(\kappa)$ by

$$R(A, B) \quad \text{iff} \quad \{\alpha \in \theta \mid f_A(\alpha) < f_B(\alpha)\} \text{ is stationary.}$$

6.2 Lemma

Let $A, B \subseteq \kappa$, $A \neq B$. Then either $R(A, B)$ or $R(B, A)$ (or both).

Proof : Clearly,

$$\theta = \{\alpha \in \theta \mid f_A(\alpha) < f_B(\alpha)\} \cup \{\alpha \in \theta \mid f_B(\alpha) < f_A(\alpha)\}$$
$$\cup \{\alpha \in \theta \mid f_A(\alpha) = f_B(\alpha)\} .$$

By our earlier discussion, there is a $\gamma < \theta$ such that

$$\{ \alpha \in \theta \mid f_A(\alpha) = f_B(\alpha)\} \subseteq \gamma.$$

Suppose neither $R(A, B)$ nor $R(B, A)$ held. Then we could find club sets $C_1, C_2 \subseteq \theta$ such that

$$\alpha \in C_1 \rightarrow f_A(\alpha) \text{ is not less than } f_B(\alpha) ,$$

$$\alpha \in C_2 \rightarrow f_B(\alpha) \text{ is not less than } f_A(\alpha) .$$

By 2.2, $C = C_1 \cap C_2 - (\gamma+1)$ is club in θ. But by our initial remark we must have $C = \emptyset$, which is a contradiction. This proves that at least one of $R(A, B)$ and $R(B, A)$ must be valid. (They could both be valid.) □

Our aim is to prove $2^\kappa = \kappa^+$. We assume, on the contrary, that $2^\kappa > \kappa^+$, and work towards a contradiction.

6.3 Lemma

There is a $B \subseteq \kappa$ such that $|\{A \subseteq \kappa \mid R(A, B)\}| \geq \kappa^+$.

Proof : Let $X \subseteq \mathcal{P}(\kappa)$, $|X| = \kappa^+$. If there is a $B \in X$ with the required property we are done, so assume otherwise. For each $B \in X$, let $R^{-1}(B)$ denote the set $\{A \subseteq \kappa \mid R(A, B)\}$. Let

$$Y = \cup\{R^{-1}(B) \mid B \in X\}.$$

Now, $|X| = \kappa^+$ and, by our assumption, $|R^{-1}(B)| \leq \kappa$ for all $B \in X$, so $|Y| \leq \kappa^+$. So, as $|\mathcal{P}(\kappa)| > \kappa^+$ there is a $B \subseteq \kappa$ such that $B \notin Y$.

Now, if $A \in X$, then $B \notin R^{-1}(A)$, so $R(B, A)$ fails. Hence, by 6.2, $A \in X$ implies $R(A, B)$. Thus as $|X| = \kappa^+$, B is as required. □

We fix B as in 6.3 from now on.

Now, for each $\alpha < \theta$, $f_B(\alpha) < \kappa_\alpha{}^+$, so we can fix some one-one mapping

$$g_\alpha : f_B(\alpha) \to \kappa_\alpha.$$

Suppose now that $A \subseteq \kappa$ is such that $R(A, B)$. Let

$$S_A = \{\alpha \in \theta \mid f_A(\alpha) < f_B(\alpha)\}.$$

Then S_A is stationary, and for each $\alpha \in S_A$, $g_\alpha \circ f_A(\alpha) < \kappa_\alpha$. So by 6.1 there is a stationary set $T_A \subseteq S_A$ and an ordinal $\gamma_A < \theta$ such that

$$\alpha \in T_A \to g_\alpha \circ f_A(\alpha) < \kappa_{\gamma_A}.$$

Now, $|\{(T_A, \gamma_A) \mid A \subseteq \kappa \ \& \ R(A, B)\}| \leq |\mathcal{P}(\theta) \times \theta| = 2^\theta.\theta = \theta^+$ (since $\theta < \kappa$).

So as there are at least κ^+ many sets $A \subseteq \kappa$ with $R(A, B)$ and κ^+ is regular, there is a pair (T, γ) such that

$$|\{A \subseteq \kappa \mid R(A, B) \ \& \ T_A = T \ \& \ \gamma_A = \gamma\}| \geq \kappa^+.$$

But $$|T_{\kappa_\gamma}| = \kappa_\gamma^{\theta} \leq \max(\kappa_\gamma^{\kappa_\gamma}, \theta^\theta)$$

$$= \max(2^{\kappa_\gamma}, 2^\theta)$$

$$= \max(\kappa_\gamma^{+}, \theta^+)$$

$$< \kappa.$$

Hence there must be sets $A_1, A_2 \subseteq \kappa$, $A_1 \neq A_2$, such that $R(A_1, B)$, $R(A_2, B)$, $T_{A_1} = T_{A_2} = T$, $\gamma_{A_1} = \gamma_{A_2} = \gamma$, and

$$g_\alpha \circ f_{A_1} \restriction T = g_\alpha \circ f_{A_2} \restriction T.$$

Since g_α is one-one, this implies

$$f_{A_1} \restriction T = f_{A_2} \restriction T .$$

But $\{\alpha \in \theta \mid f_{A_1}(\alpha) = f_{A_2}(\alpha)\}$ is known to be bounded in θ, so we have a contradiction. We have thus proved:

6.4 <u>Theorem</u>

Let κ be a singular cardinal of uncountable cofinality. If $2^\lambda = \lambda^+$ for all $\lambda < \kappa$, then $2^\kappa = \kappa^+$. □

Chapter V. The Axiom of Constructibility

1. Constructible Sets

Before reading this chapter, the reader should go back and re-read sections 1 and 2 of Chapter II, where we developed the concept of the set theoretic hierarchy, V_α, $\alpha \in On$.

Now, in defining the set theoretic hierarchy, we took as a basic notion the unrestricted power set operation $\mathcal{P}(x)$. Given the level V_α of the hierarchy, we took $V_{\alpha+1} = \mathcal{P}(V_\alpha)$: $V_{\alpha+1}$ is the set of *all* subsets of V_α. But we did not say just what does constitute a *subset* of V_α. That is, we never really defined the notion of what a set *is*! (Of course, as we said in Chapter II, a set *is* a collection of sets, but this does not tell us what a set *is* unless we know what a collection *is*.) Now, for a large part of mathematics, indeed the greatest part, this is not important. Usually in mathematics, when one needs to refer to a specific set one has a description of that set (i.e. a definition of the set), and the Axiom of Subset selection suffices. The only exception is (usually) when an Axiom of Choice or Zorn's Lemma argument is involved. Here one simply appeals to the axiom to provide a raw existence assertion. (This of course explains why some people still feel uneasy about the use of Axiom of Choice arguments: one obtains a set which one cannot "imagine".) But when we come to a question such as whether $2^{\aleph_0} = \aleph_1$ or not, the situation is quite different. Here we want to know how many elements the set $\mathcal{P}(\omega)$ has. But since we have at no point determined what is to constitute an arbitrary subset of ω, how could we expect to answer this question? It turns out that indeed we cannot. The Zermelo-Fraenkel axioms do not decide the continuum problem one way or the other. (We sketch a proof of this in Chapter VI.) Of course, one could assume that this type of question is the only type which results in an undecidable statement and ignore it, but this is not reasonable. There are many simple statements of analysis, for instance, which have an easy proof if $2^{\aleph_0} = \aleph_1$ but apparently no proof otherwise : and these questions demand an answer. To which problem a simple solution is to take CH (or even GCH) as an additional axiom of set theory. But why? What possible intuition could lead to our taking GCH as a "reasonable" assertion about sets? There is indeed none. And anyway, even if we were to take GCH as an axiom, our problems would not be over. There are several fundamental questions of pure mathematics which cannot be resolved even if we assume GCH. Let us state two such.

I. (Whitehead Problem). Suppose G is an abelian group with the property that whenever H is an abelian group extending the group, Z, of integers, such that $H/Z \simeq G$, then $H \simeq Z \oplus G$ (direct sum). Is G necessarily free? (G being free is a sufficient condition for this to hold.)

II. (Souslin Problem). Let $(X, <)$ be a Dedekind complete toset with no end points, such that between each pair of elements of X lies a third element of X. Suppose that there is no uncountable collection of pairwise disjoint open intervals of X. Is it necessarily the case that $(X, <) \cong \mathbb{R}$? (If the last condition is strengthened to X having a countable dense subset the answer is "Yes".)

Assuming that we feel that our foundational set theory should be able to provide the means of resolving such questions, we had better re-examine our set theory. We describe one natural and highly successful solution to the dilemma (which certainly resolves the two questions above, as well as a good many more). The idea is to provide a precise definition of the notion of a "set" (or "collection").

Suppose we take as our basic idea of a set the notion of a *describable collection* (of sets). We can make this a bit more precise by restricting our "descriptions" to those expressible in our formal language LAST. This will allow us to refer to existing sets in order to describe new sets, because LAST includes a facility for such references. And it will clearly provide us with all the sets we need in mathematics, except *perhaps* for the "undescribed" sets which we obtain by using the Axiom of Choice. But let us leave the problem about the Axiom of Choice for the time being. (It will turn out that this is a wise decision. This is one of the rare occasions when a problem *disappears* as a result of its being ignored!) Since our original motivation for our sets forming a hierarchy would still appear to be in order, let us now try to re-define the set theoretic hierarchy, replacing the unrestricted (and undescribed) power set operation by the more precise notion of the "describable power set" operation. Thus, we shall start with the empty set, and at limit levels we shall collect everything together just as before. But in proceeding from stage α to stage $\alpha+1$ we shall introduce just those subsets (of what we now have) which we can describe using LAST. To indicate that the hierarchy is being defined differently, we denote the α'th level not by V_α now but by L_α. Thus, we have the (tentative and as yet informal) definition :

$$L_0 = \emptyset ;$$

$$L_\lambda = \bigcup_{\alpha<\lambda} L_\alpha , \text{ if } \lim(\lambda) :$$

$L_{\alpha+1}$ = all collections of elements of L_α which are describable by means of a formula of LAST.

All we need to do now is to make the last clause in this definition precise. We will not run into any problems providing we keep in mind the fundamental intuition that, when we are trying to define $L_{\alpha+1}$, *those sets in* L_α *and only those sets are at our disposal.*

Assume then that we have constructed the set L_α. If $\phi(v_n)$ is a formula of LAST having the free variable v_n (only), and if $a_1, \ldots, a_m$ are sets in L_α which the names (i.e. the w_i's) in ϕ denote, then the collection of all those sets x in in L_α for which $\phi(x)$ is true is well-defined. $L_{\alpha+1}$ will consist of all such collections. Thus, $X \in L_{\alpha+1}$ iff there is a formula $\phi(v_n)$ of LAST with free variable v_n (only) and sets $a_1, \ldots, a_m$ in L_α which interpret the names involved in ϕ, such that X is the collection of all x in L_α for which $\phi(x)$ is true.

One point needs a little clarification here. Suppose the formula ϕ involves the qualifier $\forall v_i$. What do we mean by saying "$\phi(x)$ is true". Well, at the stage α, the only sets available are those in L_α. So we are only in a position to "check" whether all interpretations of v_i *in* L_α are as required. In other words, the only possible meaning which the quantifier $\forall v_i$ can have at stage α is "for all v_i in L_α". Similarly for an existential quantifier : $\exists v_j$ can only mean "there exists a v_j in L_α". (Strictly, *at stage* α there is no need for the qualification "in L_α" here, since L_α really is all there is!) Thus the truth or falsity of $\phi(x)$ at stage α need not be related to its eventual "truth" or "falsity". But since L_α is a well-defined set, the notion of "$\phi(x)$ being true with respect to the partial universe L_α" is certainly well-defined. Hence $L_{\alpha+1}$ is a precisely defined collection of sets.

That then defines our hierarchy. We set

$$L = \bigcup_{\alpha \in On} L_\alpha .$$

We'll call the class L the *constructible universe*. The notion of a set to which it corresponds is called a *constructible set*. The hierarchy L_α, $\alpha \in On$, is the *constructible hierarchy*. Of course, our notion of "describable collection" is very strong, so one should not read too much into the word "constructible" here. For instance, the real line turns out to have a *constructible* well-ordering in constructible set theory!

2. The Constructible Hierarchy

A brief examination of the above definitions shows that $L_\alpha \subseteq L_\beta$ for $\alpha \le \beta$, that each L_α is transitive, and that $L_\alpha \cap On = \{\beta \mid \beta < \alpha\} = \alpha$. These properties are shared by the V_α-hierarchy of course. But there the similarity ends. For example, because the language LAST is countable, $|L_\alpha| = |\alpha|$ for every infinite ordinal α ; hence, in particular $|L_{\omega+1}| = \aleph_0$. But since $\mathcal{P}(\omega) \subseteq V_{\omega+1}$, $|V_{\omega+1}| > \aleph_0$.

Thus the constructible hierarchy grows much more slowly than the Zermelo hierarchy.

And now, before we go on, let us explain a point that may just have begun to worry our reader. Does not the fact that $L_{\omega+1}$ is countable contradict the fact that $\mathcal{P}(\omega)$ is uncountable? Well, since we have not yet examined the consequences of our new notion of a set, it may be that in our new set theory $\mathcal{P}(\omega)$ is in fact countable. But before the reader throws up his hands in horror, let us hasten to say that this is not in fact the case. $\mathcal{P}(\omega)$ *is* uncountable in constructible set theory. The confusion (if there is any) lies in the fact that $\mathcal{P}(\omega)$ will not be contained in $L_{\omega+1}$. Certainly, some subsets of ω will lie in $L_{\omega+1}$. For instance, the set of all even numbers is there, as too is the set of all multiples of 3. Indeed, $L_{\omega+1}$ will contain infinitely many subsets of ω. But to be in $L_{\omega+1}$, a subset of ω will have to be describable *in terms of sets in* L_ω. This only allows the formation of relatively simple sets of numbers. L_ω does not contain enough "information" to enable us to define "complex" sets of integers. But already when we come to define $L_{\omega+2}$ our expressive power has increased enormously. We may now refer to all the new sets which went into $L_{\omega+1}$ in our descriptions. Thus $L_{\omega+2}$ will contain many new sets of integers, not "constructible" before. And so on. Thus, not only does the constructible hierarchy grow more slowly than the Zermelo hierarchy, it in fact grows in quite a different manner.

3. The Axiom of Constructibility

We are now in a position analogous to the position we were in at the end of II.3. By means of an analysis of the notion of a "set" we have arrived at a picture of how the set theoretic universe should look. Instead of the picture represented by the two "axioms"

(Z1) $\quad V = \bigcup_{\alpha \in On} V_\alpha$;

(Z2) $\quad$ Axiom of subset selection ,

we now have two principles

(L1) $V = \bigcup_{\alpha \in On} L_\alpha$;

(L2) Axiom of subset selection.

(So far in our discussion we have not mentioned (L2), but of course we shall need this if our set theory is to be of any use to us as mathematicians. And the remarks we made after defining the constructible hierarchy should indicate why it may be necessary to include this principle as an axiom, even though the constructible hierarchy is built up by defining sets; namely, defining subsets of L_α *at stage* α is not at all the same as defining subsets of L_α *over the entire universe*, which is what the Axiom of Subset selection concerns.)

The next step is to do what we did in II.4 for the Zermelo-Fraenkel set theory: to analyse the two principles (L1) and (L2) and thereby isolate all those assumptions about sets of which the construction makes implicit use.

Well, we certainly need the ordinal number system. We also need the recursion principle. (The formal definition of the constructible hierarchy will be as a recursion on ordinals, of course) In fact, the only difference between the constructible hierarchy and the Zermelo hierarchy lies in what we do at successor stages. With the Zermelo hierarchy, since the power set operation is guaranteed by the ZF axioms, the ZF system sufficed for the entire construction. But the definition of $L_{\alpha+1}$ from L_α is a little more complex. Here we use various concepts of logic : formulas, assignments of sets to names, interpretation of variables, truth of formulas within a certain partial universe L_α. Now, admittedly mathematical logic (or rather the parts of it which concern us here) deals with some of the fundamental concepts which lie behind the notion of a set. Nevertheless, mathematical logic, in common with all other areas of pure mathematics, can be developed rigorously *within set theory*. In particular, all of the concepts required for the passage from L_α to $L_{\alpha+1}$ are capable of definition and analysis within set theory. Indeed, ZF suffices! In other words, the construction of the constructible hierarchy is possible on the basis of the ZF axioms, just as was the construction of the Zermelo hierarchy. (The development of mathematical logic within set theory is not particularly difficult, but it would constitute too great a digression to go into details here. All that we need to know for our discussion is that within ZF one can define a function Def : $V \to V$ such that $\mathrm{Def}(L_\alpha) = L_{\alpha+1}$ for all α. Def(X) is the set of all "definable" subsets of X, for any set X, where "definable" means "definable over the partial universe X by means of a formula of LAST, with one free variable, whose names refer only to sets in X". So Def corresponds to $\wp$ in the definitions of the two hierarchies.)

Hence, constructible set theory can be axiomatised thus :

(A) The ZF axioms ;

(B) $V = \bigcup_{\alpha\in On} L_\alpha$.

(A) enables us to define the hierarchy L_α, $\alpha \in On$. (B) tells us that the universe of sets is the limit of this hierarchy. Consequently, we see that constructible set theory is an *extension* of ZF, obtained by adjoining the extra axiom (B). We call (B) the *Axiom of Constructibility*. Since we have introduced the symbol L to denote $\bigcup_{\alpha\in On} L_\alpha$, the axiom of constructibility may be abbreviated thus:

$$V = L .$$

And constructible set theory may be denoted as :

$$ZF + (V = L) .$$

Now, on the basis of the ZF axioms we can still define the Zermelo hierarchy, regardless of whether V = L or not. Hence, V = L does not affect the validity of the equation

$$V = \bigcup_{\alpha\in On} V_\alpha .$$

But V = L certainly does affect the meaning of this equation. In the context of ZF alone, the power set operation is left totally undescribed, which means that there is a great degree of "freedom" built in to the Zermelo hierarchy. But if we assume V = L (in addition to the ZF axioms) then the notion of what constitutes a set is very precise, which means that the power set operation is a rigidly determined operator.

And now to the Axiom of Choice. The one obvious advantage of leaving the power set unrestricted is that it allows one to postulate the existence of choice sets, and thereby to introduce the Axiom of Choice. But if we adopt as our system of set theory the theory ZF + (V = L), we no longer have this freedom. Either AC will be true or it will be false. Fortunately for us it turns out to be true:

3.1 Theorem (of the system ZF + (V = L).)

Every set can be well-ordered. □

We shall not prove 3.1, but we can give a brief indication of how the proof goes. The idea is to prove, by induction, that each set L_α can be well-ordered. (Since each set is a subset of some L_α in constructible set theory, this clearly suffices.) And we do this as follows. Clearly, L_0 can be well-ordered. And if $\lim(\alpha)$ and each L_β can be well-ordered for $\beta < \alpha$, then $L_\alpha = \bigcup_{\beta<\alpha} L_\beta$ can be well ordered by combining the well-orderings of the L_β, $\beta < \alpha$. (This is only a sketch, remember.) Now suppose L_α can be well-ordered. It is easy to define a well-ordering of the formulas of LAST which have one free variable. Thus, using the well-ordering of L_α we can define a well-ordering of all these formulas of LAST coupled with the interpretations of the names in the formulas as elements of L_α. But this in effect provides us with a well-ordering of $L_{\alpha+1}$.

In view of 3.1, we can in fact regard constructible set theory as an extension not just of ZF but of ZFC - Zermelo-Fraenkel set theory. And we have the added bonus that there is no "problem" about AC. It is provable from the other axioms. This is the way constructible set theory is currently regarded. For most applications of set theory, it is not necessary to define precisely the concept of a set. The Zermelo-Fraenkel picture of the universe suffices. So why assume more? Hence we take ZFC as the basic set theory of mathematics. This leaves some questions in mathematics unresolved. To answer these questions we need to be more precise as to what a set really is. Whether or not we regard the constructible set theory as "more natural" than the Zermelo-Fraenkel system (and many mathematicians do), if we are subsequently able to solve the problem in constructible set theory, then the effect is that by assuming an additional *axiom*, the problem can be solved. In this case what the "axiom" effectively says is that we are fixing a precise definition of the set concept. Now, if one regards constructible set theory as a reasonable theory of sets, any result proved in it will be simply a "theorem". If one does not so regard constructible set theory, its results will just be "theorems based on an additional assumption". (Some people regard AC in this manner as well.) But either way, it is worth noting that the notion of constructibility is needed for the result. Since ZFC is taken as basic, we never mention the use of the ZFC axioms (except that we sometimes mention that AC is necessary for a result). Consequently, when we prove a result in the system ZFC + (V = L), it suffices to prefix the result with the statement : "Assume V = L." This means at once that the theorem concerned is to be proved within the framework of constructible set theory and not just Zermelo-Fraenkel set theory.

4. The Consistency of Constructible Set Theory.

We have indicated earlier (see II.7) that in a theory of sets one cannot ever hope to prove (within that system) the consistency of the theory. Thus, just as we cannot prove within the system ZFC that ZFC is a consistent theory, so too are we unable to prove within the system ZFC + (V = L) that this system is consistent. Thus, if we take as our basic set theory the constructible set theory, we must simply *assume* that as a formalisation of our intuitions concerning sets it is a consistent system. But we have also seen that we can regard constructible set theory as an extension of Zermelo-Fraenkel obtained by adding an extra axiom, the axiom of constructibility. Viewed in *this* light, the constructible set theory could be said to be somewhat more "suspect" regarding its consistency than ZFC : the more axioms we have, the more chance there is that there will be an internal inconsistency. Which could be used as an argument against constructible set theory. But in fact there is no such danger, by virtue of the following theorem of Gödel :

4.1 Theorem

If ZF is a consistent theory, so too is ZF + (V = L). □

A rigorous proof of 4.1 is beyond the scope of this book. Intuitively the idea is as follows. In order to prove a system of axioms is consistent, what one must do is exhibit a "model" for that system. Starting with a model of ZF (which exists as a consequence of the assumption of 4.1 that ZF is consistent), one can define within that model the "constructible universe". This miniature "constructible universe" turns out to constitute a model of ZF + (V = L). (Incidentally, the proof of 4.1 itself takes place in a very simple fragment of ZF.) Since AC is a theorem of the theory ZF + (V = L), 4.1 at once implies the corollary :

4.2 Corollary

If ZF is a consistent theory, so too is ZFC. □

5. Use of the Axiom of Constructibility.

One of the simplest consequences of the axiom of constructibility is the solution of the continuum problem :

5.1 Theorem

Assume V = L. Then GCH holds. □

Unfortunately, even a sketch of the proof is beyond the scope of this book. This is not because the proof is particularly complex. The difficulty lies in the fact that it requires a reasonable knowledge of the techniques of mathematical logic. This is to be expected with proofs which make an essential use of the Axiom of Constructibility. *The fact that a result is not provable in ZFC alone already means that a detailed analysis of the notion of sets is required for its solution.* And such an investigation is of course a matter of mathematical logic. Now, since mathematical logic is a well defined mathematical discipline, the proofs within this field resemble proofs in any area of mathematics : they do not stand out as unusual in any way. But to follow such a proof naturally requires a degree of familiarity with the field. For instance, to prove that $2^{\aleph_0} = \aleph_1$ (assuming $V = L$), one demonstrates that, although new subsets of ω keep appearing as we proceed up the constructible hierarchy through $L_{\omega+1}$, $L_{\omega+2}$, ... etc., this process terminates by stage ω_1, so that $\mathcal{P}(\omega) \subseteq L_{\omega_1}$. Since $|L_{\omega_1}| = \aleph_1$, this implies at once that $2^{\aleph_0} = \aleph_1$. But to prove that the process of new sets of integers appearing stops at stage ω_1 requires a fairly deep analysis of the constructible hierarchy and the way it grows.

The fact that proofs involving $V = L$ involve a good knowledge of mathematical logic means, of course, that most working mathematicians are unable to work in constructible set theory. But this is not always the case. Set theorists have obtained various principles of combinatorial set theory within the system $ZFC + (V = L)$, and for many applications these consequences of $V = L$ are all that is required. For instance, one of the most common combinatorial consequences of $V = L$ is the following, known as $\diamond$:

There is a sequence $\langle S_\alpha \mid \alpha < \omega_1 \rangle$ such that for each $\alpha < \omega_1$, $S_\alpha \subseteq \alpha$, and whenever $X \subseteq \omega_1$ then for some infinite ordinal $\alpha \in \omega_1$, $X \cap \alpha = S_\alpha$.

$\diamond$ is, as we said, a consequence of $V = L$. And $\diamond$ clearly implies CH. (CH does not, however, imply $\diamond$.) Many results in set theory and topology can be proved by a fairly straightforward argument which makes use of the principle $\diamond$. The mathematical logic involved in constructibility lies in the proof of $\Diamond$. To apply $\Diamond$ one needs to know nothing of mathematical logic. For example, assuming $\Diamond$, it is quite a straightforward matter to obtain a negative answer to the Souslin Problem, stated in section 1. (The Whitehead Problem requires another, rather similar set theoretic principle, but again the argument from this principle needs no logic.) And of course, any result proved using GCH is automatically a theorem of constructible set theory, though such proofs usually don't involve any logic.

For any further details on the usage of V = L we refer the reader to our monograph [2].

Chapter VI. Independence Proofs in Set Theory

1. Some Examples of Undecidable Statements

The following statements are known to be undecidable in the system ZFC. (Though they are all decidable in constructible set theory, by the way.)

(A) The Whitehead Problem (see V.1 for a statement of the problem)

(B) The Souslin Problem (ditto)

(C) Borel's Conjecture Let $X \subseteq \mathbb{R}$ and suppose that whenever $\{\varepsilon_n\}$ is a sequence of positive reals there is a sequence $\{I_n\}$ of open intervals such that length $I_n < \varepsilon_n$ for each n and $X \subseteq \bigcup_{n=1}^{\infty} I_n$. Then X is countable.

(D) The union of fewer than $2^{\aleph_0}$ many sets of reals of (Lebesgue) measure zero has measure zero.

(E) The Continuum Hypothesis If $X \subseteq \mathbb{R}$ is not equinumerous with $\mathbb{R}$ then X is countable.

(F) There is a well-ordering of $\mathbb{R}$ which is definable in analysis.

2. The Idea of a Boolean-Valued Universe

We motivate a method by which one can prove, in ZFC, that the statements listed above are undecidable in ZFC. To do this we commence by a reexamination of the Zermelo hierarchy. Recall the basic definition:

$$V_0 = \emptyset ;$$

$$V_{\alpha+1} = \mathcal{P}(V_\alpha) ;$$

$$V_\lambda = \bigcup_{\alpha<\lambda} V_\alpha, \text{ if } \lim(\lambda).$$

Suppose now that we decide to develop our set theory using not sets themselves but rather characteristic functions of sets. Consider the following definition :

$$V_0^F = \emptyset ;$$

$$V_{\alpha+1}^F = {}^{(V_\alpha^F)}2 ;$$

$$V_\lambda^F = \bigcup_{\alpha<\lambda} V_\alpha^F , \text{ if } \lim(\lambda).$$

In passing from V_α^F to $V_{\alpha+1}^F$ we take not $\mathcal{P}(V_\alpha^F)$ but the set of all the characteristic functions of the members of $\mathcal{P}(V_\alpha^F)$. In essence, we will just obtain a functional equivalent of the Zermelo hierarchy. The correspondence is not quite trivial, of course, because there will be many different functions in the V_α^F-hierarchy which correspond to each set in the V_α-hierarchy: for instance, if $x \subseteq V_\alpha^F$ and $f : V_\alpha^F \to 2$ is the characteristic function of x, then $f' \in V_{\alpha+2}^F$ also corresponds to x, where

$$f' = f \cup \{(a, 0) \mid a \in V_{\alpha+1}^F - V_\alpha^F\} .$$

(The point being that when functions are involved, different domains mean different functions, even though the different functions may be "essentially" the same.) But discounting this minor technical problem, the two hierarchies V_α, $\alpha \in On$, and V_α^F, $\alpha \in On$, are essentially equivalent. Setting $V^F = \bigcup_{\alpha\in On} V_\alpha^F$ we obtain a universe of characteristic functions of "sets". (We write "sets" in quotation here because, of course, each function in V^F is itself defined on functions and not on sets. So in V^F there is really only one kind of animal : a characteristic function.) It is intuitively clear that anything we could do with V we can do with V^F. In other words, we could carry out our entire set theoretic development using the members of

V^F instead of the pure sets of V. (Few readers would regard this as a worthy exercise, of course, and we are not for one moment suggesting that it should be done. But it is certainly possible.)

And now we ask ourselves what the significance is of the fact that we only allow functions mapping into 2? Well, the elements 1, 0 of 2 correspond to the two truth values, T and F ("true" and "false", respectively). If $f \in V^F$ and $f(x) = 1$, then the statement "$x \in f$" (interpreted in V^F) is true, i.e. has the truth value T. And if $f(y) = 0$, the statement "$y \in f$" (interpreted in V^F) has the truth value F. Hence, our restriction to functions mapping into 2 corresponds to the fact that our logic admits only two possibilities, true or false. But why not allow more possibilities? Certainly we are all aware that in real life there are more than just two truth values, as the following anecdote of P. Vopenka illustrates. According to Charles Darwin†, there is a finite toset whose first element is a monkey and whose last element is you, dear reader. Let $M(x)$ denote the statement that "x is a monkey ". Let x_0 be the first member of our finite toset, x_1 the second, and so on, with you being x_n. By assumption, $M(x_0)$. In two valued logic, we clearly have $M(x_m) \rightarrow M(x_{m+1})$ for any m. (The offspring of a monkey is a monkey). Hence, by a simple induction we conclude that $M(x_n)$! Assuming you agree that we have now arrived at a contradiction, let us see what has gone wrong. Well, nothing really, except that it is not valid to use two valid logic here. Although $M(x_0)$ holds and $M(x_n)$ fails, in between there is a gradual change in truth values, with $M(x_m)$ becoming "less true" as m increases

Of course, the above anecdote does not in itself constitute a sufficient reason for adopting a many valued logic in mathematics. But it does illustrate that such a concept is not entirely devoid of meaning. And it turns out that this is the device which we need in order to obtain our undecidability results.

So what sort of sets can we replace 2 by and still obtain a "universe of sets" which has some useful properties? Besides the fact that it corresponds to our

---

†Unfortunately, this particular example only works for those who accept the Darwinian theory of evolution. But it is easy to construct other examples.

intuition concerning *two* truth values, what is so special about the set {0, 1}? The answer is that it does correspond to *truth values*. For instance, in V^F if $f : V_\alpha^F \to 2$ and $g : V_\alpha^F \to 2$ is defined by $g = 1 - f$, then we have for any $x \in V_\alpha^F$, $f(x) = 1$ iff $g(x) = 0$. And if we set $h = \min(f, g)$, then $h(x) = 1$ iff $f(x) = 1$ and $g(x) = 1$. And so on. If our functional hierarchy is to provide us with a type of "set theory", then the values of the functions must behave like truth values. Well, what kinds of set do behave like truth values? The answer is well known. Boolean algebras! (See Problem I.1 for relevant definitions.) Providing $\mathbb{B}$ is a boolean algebra, we obtain a reasonable "universe of sets" by means of the definition :

$$V_0^{\mathbb{B}} = \emptyset ;$$

$$V_{\alpha+1}^{\mathbb{B}} = \{f \mid f : V_\alpha^{\mathbb{B}} \to \mathbb{B}\} ;$$

$$V_\lambda^{\mathbb{B}} = \bigcup_{\alpha<\lambda} V_\alpha^{\mathbb{B}} , \text{ if } \lim(\lambda) ;$$

$$V^{\mathbb{B}} = \bigcup_{\alpha \in On} V_\alpha^{\mathbb{B}} .$$

An element of $V^{\mathbb{B}}$ is called a *boolean-valued set*, or more precisely a $\mathbb{B}$-*valued set*. $V^{\mathbb{B}}$ is a *boolean-valued universe*, or more precisely *the* $\mathbb{B}$-*valued universe*. If $x \in V_\alpha^{\mathbb{B}}$ and $f \in V_{\alpha+1}^{\mathbb{B}}$, $f(x)$, which is an element of $\mathbb{B}$, is a measure of the truth of the statement "$x \in f$" in terms of $V^{\mathbb{B}}$. If $f(x) = \mathbb{0}$, then x is certainly not a member of f; if $f(x) = \mathbb{1}$, x is a member of f; and if $\mathbb{0} < f(x) < \mathbb{1}$, then x is partly not in f and partly in f, with f(x) telling us "to what extent" x is a member of f.

3. <u>The Boolean-Valued Universe</u>

We now formalise the discussions of section 2. For technical reasons we shall set things up in a slightly different manner. For a start, we shall not use any old boolean algebra $\mathbb{B}$, but rather a *complete* boolean algebra. (A boolean algebra $\mathbb{B}$ is complete iff every subset, X, of $\mathbb{B}$ has a least upper bound — denoted by $\bigvee X$ — and a greatest lower bound - denoted by $\bigwedge X$.) Secondly, we shall not demand that our $\mathbb{B}$-valued characteristic functions are defined on some $V_\alpha^{\mathbb{B}}$: they can have arbitrary domains. (Since there will in any case be a great deal of duplication, with many members of $V^{\mathbb{B}}$ denoting the same "set", owing to differing domains, this causes no extra hardship, and simplifies matters a little.)

So fix now some complete boolean algebra $\mathbb{B}$. By recursion on ordinals we define the hierarchy of $\mathbb{B}$-*valued sets* as follows :

$$V_\alpha^{\mathbb{B}} = \{u \mid u \text{ is a function } \& \text{ ran}(u) \subseteq \mathbb{B} \ \& \ (\exists\beta < \alpha)(\text{dom}(u) \subseteq V_\beta^{\mathbb{B}})\}.$$

(This formulation allows for the cases $\alpha = 0$, succ(α), lim(α) in one go, but is easily seen to be equivalent to taking :

$$V_0^{\mathbb{B}} = \emptyset ;$$

$$V_{\alpha+1}^{\mathbb{B}} = \{u \mid \text{dom}(u) \subseteq V_\alpha^{\mathbb{B}} \ \& \ \text{ran}(u) \subseteq \mathbb{B}\} ;$$

$$V_\lambda^{\mathbb{B}} = \bigcup_{\alpha<\lambda} V_\alpha^{\mathbb{B}}, \text{ if } \lim(\lambda).\ )$$

Clearly, $\alpha < \beta \rightarrow V_\alpha^{\mathbb{B}} \subseteq V_\beta^{\mathbb{B}}$. We set

$$V^{\mathbb{B}} = \bigcup_{\alpha\in On} V_\alpha^{\mathbb{B}}.$$

$V^{\mathbb{B}}$ is the $\mathbb{B}$-*valued universe*. The elements of $V^{\mathbb{B}}$ are $\mathbb{B}$-*valued sets*. (Thus a $\mathbb{B}$-valued set is a $\mathbb{B}$-valued function defined on $\mathbb{B}$-valued sets.)

Having defined the $\mathbb{B}$-valued universe, we must assign $\mathbb{B}$-truth values to the various set theoretical assertions we can make about the members of $\mathbb{B}$. For each sentence ϕ of LAST, providing we know which "sets" in $V^{\mathbb{B}}$ the names in ϕ refer to, we should be able to assign to ϕ a unique "truth value" which measures the degree to which ϕ is true. We shall denote this truth value by

$$\| \phi \| .$$

Thus $\|\phi\|$ will always be a member of $\mathbb{B}$. If $\|\phi\| = \mathbb{0}$, ϕ will be *false* in $V^{\mathbb{B}}$. If $\|\phi\| = \mathbb{1}$, ϕ will be *true* in $V^{\mathbb{B}}$. In all other cases, ϕ will be partly false and partly true in $V^{\mathbb{B}}$. The definition of $\|\phi\|$ is obtained by unravelling the construction of ϕ. We consider first the case where ϕ is an "atomic" sentence of the form $w_i \in w_j$ or $w_i = w_j$. To avoid talking of "names" and their "meanings", we shall henceforth just use x, y, z, u, v, w, etc. to denote both names and their meanings. This accords with common usage both in and out of logic.

If $u, v \in V^{\mathbb{B}}$ then, how should we define

$$\|u \in v\| \quad , \quad \|u = v\| \quad ?$$

Well, intuitively, $v(u)$ measures the degree to which u is an element of v, so why not take as our definition $\|u \in v\| = v(u)$? Well, because this only works when $u \in \mathrm{dom}(v)$, whereas we want $\|u \in v\|$ to have a meaning for all u, v in $V^{\mathbb{B}}$. A similar difficulty arises with $\|u = v\|$, which needs to be defined even if $\mathrm{dom}(u) \neq \mathrm{dom}(v)$. To overcome this difficulty, we recall the following extensionality principles:

$$u \in v \quad \leftrightarrow \quad (\exists y \in v)(u = y) \;;$$

$$u = v \quad \leftrightarrow \quad (\forall x \in u)(x \in v) \wedge (\forall y \in v)(y \in u) \;.$$

Accordingly we make the definitions :

$$\|u \in v\| \quad = \bigvee_{y \in \mathrm{dom}(v)} [v(y) \wedge \|u = y\|] \;;$$

$$\|u = v\| \quad = \bigwedge_{x \in \mathrm{dom}(u)} [u(x) \Rightarrow \|x \in v\|] \wedge \bigwedge_{y \in \mathrm{dom}(v)} [v(y) \Rightarrow \|y \in u\|] \;,$$

where for $b, c \in \mathbb{B}$, the element $b \Rightarrow c$ of $\mathbb{B}$ is defined by

$$b \Rightarrow c \quad = \quad -b \vee c \;.$$

By recalling the connection between the boolean operations $\wedge$, $\vee$, $-$ and their logical counterparts $\wedge$, $\vee$, $\neg$ (respectively), parts of the above definition make sense. And it is easy to see why " $\bigvee_{y \in \mathrm{dom}(v)}$ " corresponds to "$(\exists y \in \mathrm{dom}(v))$" and " $\bigwedge_{x \in \mathrm{dom}(u)}$ " to "$(\forall x \in \mathrm{dom}(u))$". But the definition will no doubt still seem a little odd. Unfortunately, to try to clarify the definition would involve so great a digression that we shall leave the matter with the remark that this is the best definition which does all we want of it. (But notice that the definition is by a "double recursion". We define $\|u \in v\|$ and $\|u = v\|$ simultaneously. In order to calculate $\|u \in v\|$ we need to know all the values of $\|u = y\|$ for $y \in \mathrm{dom}(v)$. And in order to calculate $\|u = v\|$ we need all the values of $\|x \in v\|$ for $x \in \mathrm{dom}(u)$ and all the values of $\|y \in u\|$ for $y \in \mathrm{dom}(v)$. It can be shown that this does provide us with a sound recursive definition.)

The assignments of $\mathbb{B}$-truth values to compound sentences is now quite straightforward. The conditions for the recursion are:

$$\|\phi \vee \psi\| = \|\phi\| \vee \|\psi\| ;$$

$$\|\phi \wedge \psi\| = \|\phi\| \wedge \|\psi\| ;$$

$$\|\neg \phi\| = -\|\phi\| ;$$

(Thus, for example $\|\phi \rightarrow \psi\| = \|\phi\| \Rightarrow \|\psi\|$.)

$$\|\exists u \phi(u)\| = \bigvee_{u \in V^{\mathbb{B}}} \|\phi(u)\| ;$$

$$\|\forall u \phi(u)\| = \bigwedge_{u \in V^{\mathbb{B}}} \|\phi(u)\| .$$

Notice the duplication in notation here, with the symbols $\vee$ and $\wedge$ being used in two different ways. By these very definitions, there is of course no harm in this clash, and indeed it helps to highlight the reason why we need to have a boolean algebra for our set of "truth values". (The last two clauses also indicate why the algebra should be complete, as did the definitions of $\| u \in v \|$ and $\| u = v \|$.)

4. $V^{\mathbb{B}}$ and V .

Now, all of our development so far has taken place within the framework of ZFC. ($V^{\mathbb{B}}$ is, after all, just the result of a simple set theoretic construction by recursion.) Hence $V^{\mathbb{B}}$ is a well-defined class within V:

$$V^{\mathbb{B}} \subseteq V.$$

But in a sense, $V^{\mathbb{B}}$ is an "extension" of V. For consider the particular boolean-valued universe $V^{\mathfrak{2}}$, where $\mathfrak{2}$ is the two element algebra $\{\mathbb{0}, \mathbb{1}\}$. Clearly $V^{\mathfrak{2}}$ should be "isomorphic" to V in some sense. In fact, if we define a "relation" ~ on $V^{\mathfrak{2}}$ by

$$u \sim v \quad \text{iff} \quad \|u = v\| = \mathbb{1},$$

then ~ is an "equivalence relation", and if we "factor out" $V^{\mathfrak{2}}$ by the "relation" ~ we do obtain an isomorph to V. (The quotation marks are necessary because we are dealing with proper classes here, in a manner which is not permitted within ZFC. An equivalent argument can be formulated within the ZFC framework, but it is a little more complicated.) Now, since $\mathfrak{2}$ is a complete subalgebra of $\mathbb{B}$, it is easily seen that:

(i) $V^{\mathfrak{2}} \subseteq V^{\mathbb{B}}$;

(ii) if $u, v \in V^{\mathfrak{2}}$, then: $\|u \in v\|^{\mathfrak{2}} = \|u \in v\|^{\mathbb{B}}$;

$$\|u = v\|^{\mathfrak{2}} = \|u = v\|^{\mathbb{B}}.$$

Hence $V^{\mathfrak{2}}$ is an isomorphic copy of V sitting inside $V^{\mathbb{B}}$. This is the sense in which $V^{\mathbb{B}}$ "extends" V.

In fact, there is a canonical embedding of V into $V^{\mathbb{B}}$ which is often useful. By recursion we define $\vee : V \to V^{\mathbb{B}}$ by

$$\operatorname{dom}(\check{x}) = \{\check{y} \mid y \in x\};$$

$$\check{x}(a) = \mathbb{1} \quad \text{for all } a \in \operatorname{dom}(\check{x}).$$

(In other words, $\check{x} = \{(\check{y}, \mathbb{1}) \mid y \in x\}$.) Then, for $x, y \in V$,

$$x = y \quad \text{iff} \quad \|\check{x} = \check{y}\|^{\mathbb{B}} = \mathbb{1};$$

$$x \in y \quad \text{iff} \quad \|\check{x} \in \check{y}\|^{\mathbb{B}} = \mathbb{1}.$$

There is one great source of difficulty for the beginner concerning the use of the symbols =, ϵ, etc. On the one hand, we ourselves carry out all our arguments in regular ZFC set theory, where a set is a set! On the other hand, some of our arguments involve the internal properties of the universe $V^{\mathbb{B}}$, where all "sets" are $\mathbb{B}$-valued sets. Let us stress that we as mathematicians continue to use regular set theory and logic. *Within which framework* we discuss boolean-valued sets and boolean-valued logic. Unfortunately, only experience can really overcome the problems which arise from the above situation.

5. Boolean-Valued Sets and Independence Proofs.

We shall wish to consider $\mathbb{B}$-valued arguments within the universe $V^{\mathbb{B}}$. Accordingly, we need to know that the usual rules of logic are valid in the $\mathbb{B}$-valued case. That they are is quite easily proved, but we content ourselves here with a simple statement of the result.

5.1 Lemma

(i) All the rules and axioms of propositional logic are $\mathbb{B}$-valid.

(ii) All the rules and axioms of first order logic are $\mathbb{B}$-valid.

(iii) All the axioms of equality are $\mathbb{B}$-valid. □

Let us remark that (i) was known to Boole, and is an immediate consequence of the definition of a boolean algebra; (ii) was proved by Sikorski, and, like (i), has nothing to do with $V^{\mathbb{B}}$ in parciular; (iii) depends upon the definition of $\|u = v\|$ we made for $V^{\mathbb{B}}$.

The following theorem (which is proved within ZFC set theory, as are all our theorems about $V^{\mathbb{B}}$) is non-trivial, and is the key part of our method for obtaining independence results.

5.2 Theorem

If ϕ is an axiom of ZFC, then $\|\phi\| = \mathbb{1}$. □

As a corollary of 5.1 and 5.2, we have at once:

5.3 Theorem

If ϕ is a theorem of ZFC, then $\|\phi\| = \mathbb{1}$. □

Suppose now that we wish to prove that a certain statement Φ is undecidable in the theory ZFC. Here is one way we might try to do this. The algebraic structure of a complete boolean algebra $\mathbb{B}$ has a considerable effect upon the structure of the universe $V^{\mathbb{B}}$. (This is a fact which I both know and appreciate. Now you know it. Unfortunately space does not permit me to help you appreciate it.) Suppose that by examination of the statement Φ we are able to find (or construct) an algebra $\mathbb{B}$ such that, when interpreted in $V^{\mathbb{B}}$, we get

$$\mathbb{0} < \|\Phi\| < \mathbb{1}.$$

By 5.3, it will follow that Φ is not a theorem of ZFC. But since $\|\Phi\| > \mathbb{0}$, $\|\neg\Phi\| = \quad - \|\Phi\| < \mathbb{1}$, so $\neg\Phi$ is likewise not a theorem of ZFC. Hence Φ is shown to be undecidable in ZFC.

This, briefly, outlines the most common method for proving undecidability results for ZFC. Since $V^{\mathbb{B}}$ is a sort of boolean-valued "model" of the system ZFC, we often refer to the method as the method of "boolean-valued models of set theory". (This method has a model-theoretic analogue where there is no explicit use of boolean-valued logic : the method is then referred to as the method of "forcing".) Once the basic theory is known, any specific independence proof thus takes the following form:

I. Examine the statement, Φ, whose independence is suspected.

II. Find or construct an algebra $\mathbb{B}$ which might do the trick.

III. Calculate $\|\Phi\|$ in $V^{\mathbb{B}}$ and see that it is neither $\mathbb{0}$ nor $\mathbb{1}$.

Each of steps I and II can involve an enormous amount of effort. Very often one is forced to adopt a different procedure:

II'. Find two algebras $\mathbb{B}_1$ and $\mathbb{B}_2$ "related" to Φ.

III'. Show that $\|\Phi\|^{\mathbb{B}_1} < \mathbb{1}$ and $\|\neg\Phi\|^{\mathbb{B}_2} < \mathbb{1}$.

(This clearly also suffices to establish the undecidability of Φ.)

Thus, although one is ultimately proving that some statement is unprovable, what is actually involved in an independence proof is a <u>proof</u> in classical set theory (where the underlining indicates that we are using the word proof in the sense the reader is no doubt used to).

6. The Non-Provability of CH†

As an example of how the above ideas are applied we sketch a proof of the fact that the CH is not provable in ZFC. This is perhaps the easiest of all independence proofs. (It was also the first.)

We commence by defining our boolean algebra. Now, for most independence proofs there is no "standard" algebra which suffices. One has to use one's "appreciation" of the statement whose undecidability is to be shown in order to construct a very special algebra which will work. (Such constructions can be very delicate and occasionally stretch into fifty pages or so.) But for CH a "standard" algebra suffices.

Let $X = 2^{\omega\times\omega_2}$, a generalised Cantor space. (That is, let 2 have the discrete

---

†We should warn the reader that this section assumes a considerable acquaintance with boolean algebras and a little measure theory. Moreover, even with the necessary prerequisites, the reader should not expect to gain more than a general impression of the proof. We do not strive for completeness in our account, and several tricky points are glossed over without mention. For a rigorous account the reader should consult, for example, the book [1] of Bell.

topology and give X the product topology induced from 2.) Let $\mathfrak{G}$ be the field of all Borel subsets of X. $\mathfrak{G}$ is a σ-field, of course. Now make X into a measure space by taking the usual measure on 2 and forming the product measure on X. Let Δ be the σ-ideal of all Borel sets of measure zero. Let $\mathbb{B} = \mathfrak{G}/\Delta$, the quotient algebra. It can be shown that $\mathbb{B}$ is complete. (In fact, the measure on X induces a measure on $\mathbb{B}$, so completeness is almost a triviality.)

The non-provability of CH follows from the fact that

$$\|2^{\aleph_0} > \aleph_1\|^{\mathbb{B}} > \mathbb{0} .$$

(Hence $\|CH\|^{\mathbb{B}} < \mathbb{1}$.) We sketch a proof of this fact.

Now, ω_1 is, by definition, the first uncountable ordinal. Hence, by isomorphism,

$$\|\check{\omega}_1 \text{ is the first uncountable ordinal}\|^{2} = \mathbb{1} .$$

But $V^{\mathbb{B}}$ contains many more "sets" than does V^{2} . And perhaps amongst these extra "sets" is one which is (in $V^{\mathbb{B}}$ terms) a map of $\check{\omega}$ onto $\check{\omega}_1$. Thus, it is possible that

$$\|\check{\omega}_1 \text{ is countable}\|^{\mathbb{B}} = \mathbb{1} .$$

Or to put it another way, we may have

$$\|\check{\omega}_1 < \omega_1\|^{\mathbb{B}} = \mathbb{1} .$$

(Where ω_1 without the hat means the ω_1 of $V^{\mathbb{B}}$, the first uncountable ordinal in the universe $V^{\mathbb{B}}$. Which is not the same as $\check{\omega}_1$, this just being the image under the embedding $\vee : V \to V^{\mathbb{B}}$ of the first uncountable ordinal in V.) Indeed, for many algebras $\mathbb{B}$ the above situation does arise. But in the present situation it does not. This follows from the fact that, being a measure algebra, $\mathbb{B}$ satisfies the countable chain condition.

6.1 <u>Lemma</u>

$$\|\check{\omega}_1 = \omega_1\|^{\mathbb{B}} = \mathbb{1} .$$

Proof : Suppose not. Thus $\|\check{\omega}_1 < \omega_1\|^{\mathbb{B}} > \mathbb{0}$. (Since $V^{\mathbb{B}}$ is "bigger" than V^{2}, it cannot happen that $\|\omega_1 < \check{\omega}_1\|^{\mathbb{B}} > \mathbb{0}$.) Hence

$$\|(\exists f)(f : \omega \xrightarrow{\text{ONTO}} \check{\omega}_1)\| > \mathbb{0}$$

Now, by induction on $n \in \omega$, $\|\check{n}$ is the n'th natural number$\| = \mathbb{1}$, so we clearly have

$$\|\check{\omega} = \omega\| = \mathbb{1} .$$

Hence the above can be written

$$\|(\exists f)(f : \check{\omega} \xrightarrow{\text{ONTO}} \check{\omega}_1)\| > \mathbb{0}.$$

So, for some $f \in V^{\mathbb{B}}$,

$$b = \|f : \check{\omega} \xrightarrow{\text{ONTO}} \check{\omega}_1\| > \mathbb{0}.$$

Then, $$b \leq \|(\forall \alpha \in \check{\omega}_1)(\exists n \in \check{\omega})(f(n) = \alpha)\| ,$$

so $$b \leq \bigwedge_{\alpha \in \omega_1} \bigvee_{n \in \omega} \|f(\check{n}) = \check{\alpha}\| .$$

So, for each $\alpha \in \omega_1$ we can pick an $n(\alpha) \in \omega$ such that

$$b \wedge \|f(\check{n}) = \check{\alpha}\| > \mathbb{0}.$$

Since ω_1 is uncountable, we can find an uncountable set $X \subseteq \omega_1$ such that $n(\alpha) = n$ (say) for all $\alpha \in X$. For each $\alpha \in X$, set

$$b_\alpha = b \wedge \|f(\check{n}) = \check{\alpha}\| .$$

Thus $b_\alpha > \mathbb{0}$. But if $\alpha \neq \beta$ are elements of X, then

$$b_\alpha \wedge b_\beta = b \wedge \|f(\check{n}) = \check{\alpha}\| \wedge \|f(\check{n}) = \check{\beta}\|$$

$$\leq \|\check{\alpha} = \check{\beta}\|$$

$$= \mathbb{0}.$$

Hence $\{b_\alpha \mid \alpha \in X\}$ are pairwise disjoint, contrary to the countable chain condition for $\mathbb{B}$. □

The above proof should indicate how it can be that the set theory of $V^{\mathbb{B}}$ is effected by the algebraic properties of $\mathbb{B}$.

A similar proof now yields (using 6.1)

6.2 <u>Lemma</u>

$$\|\check{\omega}_2 = \omega_2\|^{\mathbb{B}} = \mathbb{1} . \qquad \square$$

For $\alpha < \omega_2$ now, define functions $u_\alpha : \mathrm{dom}(\check{\omega}) \to \mathbb{B}$ by

$$u_\alpha(\check{n}) = \{p \in X \mid p(n, \alpha) = 1\}/\Delta.$$

Clearly, $$\|u_\alpha \subseteq \check{\omega}\| = \mathbb{1}.$$

Moreover, a simple (?) calculation shows that

$$\|u_\alpha = u_\beta\| = \bigwedge_{n \in \omega} [(u_\alpha(\check{n}) \Rightarrow u_\beta(\check{n})) \wedge (u_\beta(\check{n}) \Rightarrow u_\alpha(\check{n}))]$$

$$= \{p \in X \mid (\forall n \in \omega)(p(n, \alpha) = p(n, \beta))\}/\Delta.$$

(This calculation, though quite straightforward as such arguments go, does require a considerable facility with the definitions of boolean-valued truth, so the reader is not urged to try to reconstruct it, unless he really feels he needs to.)

Suppose now that $\alpha < \beta < \omega_2$. Set

$$S = \{p \in X \mid (\forall n \in \omega)(p(n, \alpha) = p(n, \beta))\}.$$

Let $n_1, \ldots, n_k \in \omega$, and set $\{p_1, \ldots, p_{2^k}\} = {}^{\{n_1,\ldots,n_k\}}2$. For $\ell = 1, \ldots, 2^k$, let

$$U_\ell = \{p \in X \mid p(n_1, \alpha) = p(n_1, \beta) = p_\ell(n_1) \ \& \ \dots$$

$$\dots \ \& \ p(n_k, \alpha) = p(n_k, \beta) = p_\ell(n_k)\}.$$

Clearly, $S \subseteq U_1 \cup \dots \cup U_{2^k}$.

But, if μ denotes the measure on X, we have

$$\mu(U_\ell) = (\tfrac{1}{2})^{2k}.$$

Hence, $\mu(S) \le 2^k . (\tfrac{1}{2})^{2k} = (\tfrac{1}{2})^k.$

So, as k is arbitrary, $\mu(S) = 0.$

Thus, $\|u_\alpha = u_\beta\| = S/\Delta = \mathbb{0}.$

Hence, in $V^{\mathbb{B}}$, the sets u_α, $\alpha < \omega_2$ are distinct subsets of ω. But $\check{\omega}_2$ is the ω_2 of $V^{\mathbb{B}}$ (by 6.2). It follows that

$$\| \, |\mathcal{P}(\omega)| \ge \aleph_2 \, \|^{\mathbb{B}} = \mathbb{1}.$$

This completes the proof that CH is not provable in ZFC. (At least, it completes our sketch of the proof. To fill in all the details entails a considerable amount of work. The interested reader should consult Bell's book [1] for more details.)

Bibliography

[1] J. L. Bell. Boolean-Valued Models and Independence Proofs in Set Theory. Oxford (1977).

[2] K. J. Devlin. The Axiom of Constructibility. Springer Lecture Notes in Mathematics 617 (1977).

[3] P. R. Halmos. Naive Set Theory. Van Nostrand (1960).

[4] P. R. Halmos. Lectures on Boolean Algebras. Van Nostrand (1963).

[5] J. D. Monk. Introduction to Set Theory. McGraw-Hill (1969).

[6] B. Rotman and G. T. Kneebone. The Theory of Sets and Transfinite Numbers. Oldbourne (1966).

Glossary of Notation

Symbol	Page
$\in$	2, 37
$\notin$	2
$\rightarrow$	3, 40
$\leftrightarrow$	3, 40
$\neg$	3, 37
$\wedge$	3, 37
$\vee$	3, 37
$\forall$	3, 38
$\exists$	3, 38
$\subseteq$	4, 41
$\subset$	4
$\nsubseteq$	4
$x \cup y$	5, 41
$x \cap y$	6
$-$	6
$\emptyset$	7
$\{a_1,\ldots,a_n\}$	8, 41
$\{ a \mid P(a) \}$	9
(a,b)	9, 41
$(a_1,\ldots,a_n)$	10
$\mathcal{P}(x)$	10
$\cup x$	10, 41
$\cap x$	11
$\bigcup_{i\in I} x_i$	11
$\bigcap_{i\in I} x_i$	11
$x \times y$	12
$x_1 \times \ldots \times x_n$	12
$[a]$	14
$\{a_n\}$	15
dom(R)	16
ran(R)	16
$f : x \rightarrow y$	17
id_x	17
$g \circ f$	17
$f[u]$	17
$f^{-1}[v]$	18
$f \restriction u$	18
$f : x \leftrightarrow y$	18
f^{-1}	19
$\bigtimes_{i\in I} x_i$	19
x^I	19

Index

Aspects of Constructibility
K. J. Devlin

"A formidable collection. . . . The presentation . . . is close to the original work of Jensen, but with some useful extra comments; much of it not available in print elsewhere. . . . It is clear that these notes are of great value . . ."
—*Mathematical Reviews*

1973/xii, 240 pp./Paper
(Lecture Notes in Mathematics, Volume 354)
ISBN 0-387-06522-9

The Souslin Problem
K. J. Devlin and **H. Johnsbraten**

In this book, the authors prove the undecidability of the classical Souslin problem (concerning the unique characterization of the real numbers by a set of canonical conditions) within the Zermelo-Fraenkel axiom system with continuum hypothesis. Most of the methods and results have been developed mainly by Jensen and are published here for the first time.

1974/viii, 132 pp./Paper
(Lecture Notes in Mathematics, Volume 405)
ISBN 0-387-06860-0

Basic Set Theory
A. Levy

From Cantor's discoveries in the 1870s to results obtained through 1975, the traditional areas of set theory spanning the last 100 years are examined. The subject is developed from the axioms in considerable detail and includes well-founded relations • cardinal arithmetic with and without the axiom of choice • definability in set theory • the theory of cardinal exponentiation (including the results of Silver and others) • Boolean algebra and Martin's axiom • weakly compact cardinals • ineffable cardinals • measurable cardinals. A thorough reading of **Basic Set Theory** will prepare the reader for further study of such topics as constructible sets, forcing, and large cardinals.

1979/350 pp./2 Tables/Cloth
(Perspectives in Mathematical Logic)